ASTROBIOLOGY AND THE SEARCH FOR LIFE IN THE UNIVERSE

HEARING

BEFORE THE

COMMITTEE ON SCIENCE, SPACE, AND TECHNOLOGY
HOUSE OF REPRESENTATIVES

ONE HUNDRED THIRTEENTH CONGRESS

SECOND SESSION

MAY 21, 2014

Serial No. 113–76

Printed for the use of the Committee on Science, Space, and Technology

U.S. GOVERNMENT PRINTING OFFICE

88–146PDF WASHINGTON : 2014

For sale by the Superintendent of Documents, U.S. Government Printing Office
Internet: bookstore.gpo.gov   Phone: toll free (866) 512–1800; DC area (202) 512–1800
Fax: (202) 512–2104   Mail: Stop IDCC, Washington, DC 20402–0001

CONTENTS

May 21, 2014

ASTROBIOLOGY AND THE SEARCH FOR LIFE IN THE UNIVERSE

WEDNESDAY, MAY 21, 2014

House of Representatives,
Committee on Science, Space, and Technology,
Washington, D.C.

The Committee met, pursuant to call, at 10:00 a.m., in Room 2318 of the Rayburn House Office Building, Hon. Lamar Smith [Chairman of the Committee] presiding.

LAMAR S. SMITH, Texas
CHAIRMAN

EDDIE BERNICE JOHNSON, Texas
RANKING MEMBER

Congress of the United States
House of Representatives
COMMITTEE ON SCIENCE, SPACE, AND TECHNOLOGY

2321 RAYBURN HOUSE OFFICE BUILDING

WASHINGTON, DC 20515-6301

(202) 225-6371

www.science.house.gov

Astrobiology and the Search for Life in the Universe

Wednesday, May 21, 2014
10:00 a.m. to 12:00 p.m.
2318 Rayburn House Office Building

Witnesses

Dr. Seth Shostak, *Senior Astronomer at the SETI Institute*

Mr. Dan Werthimer, *Director of the SETI Research Center at the University of California, Berkeley*

**U.S. HOUSE OF REPRESENTATIVES
COMMITTEE ON SCIENCE, SPACE, AND TECHNOLOGY
FULL COMMITTEE**

Astrobiology and the Search for Life in the Universe

Wednesday, May 21, 2014
10 a.m. – 11:30 p.m.
2318 Rayburn House Office Building

Purpose

The purpose of this hearing is to review the current state of the science related to the search for life in the universe.

Witnesses

- **Dr. Seth Shostak**, Senior Astronomer, SETI Institute
- **Mr. Dan Werthimer**, Director of SETI Research at the University of California Berkeley

Background

Discoveries made by the Kepler space telescope of more than 1,700 planets within the Milky Way galaxy renewed interest in the search for life in the universe.

In December 2013, the Committee held a hearing titled *Astrobiology: Search for Biosignatures in our Solar System and Beyond.* Witnesses described the different methods astrobiologists use to search for microbial life, including the study of extremophiles on Earth and the search for biosignatures in the atmospheres of planets.

This hearing continues the investigation into scientific methods being employed in the search for life in the universe. Specifically, the hearing review the state of the science associated with radio and optical astronomy.

Radio Astronomy

Radio astronomy studies the radio frequencies of celestial bodies. Astronomical phenomenon, such as stars, galaxies, pulsars and quasars, emit radio waves of varying lengths. Additionally, the cosmic background, or space in between celestial bodies, emits microwave radio waves. Radio telescopes detect these different frequencies, and astronomers use this data to characterize bodies and take scientific measurements used to understand the formation and expansion of the universe.

Radio astronomy studies the radio frequencies of celestial bodies. Astronomical phenomenon, such as stars, galaxies, pulsars and quasars, emit radio waves of varying lengths. Additionally, the cosmic background, or space in between celestial bodies, emits microwave radiation. Radio telescopes detect these different frequencies, and astronomers use this data to characterize bodies and take scientific measurements used to understand the formation and expansion of the universe.

To search for emitted signals, scientists conduct either targeted searches or sky surveys. Targeted searches are longer searches in a fixed location. Sky surveys are brief sweeps of the entire sky.

Natural radio frequencies can sometimes suffer from interference by manmade satellites and spacecraft. Astronomers must be able to differentiate between frequency sources.

Radio telescopes can be found around the globe. Some of the best known telescopes include the Atacama Large Millimeter Array in Chile, the Very Large Array in Mexico, the Arecibo telescope in Puerto Rico, the South Pole Telescope in Antarctica, and the Allen Telescope Array in northern California.

Optical Astronomy

Optical or visible-light astronomy uses a variety of light sensitive telescopes to find specific celestial bodies. Some telescopes take a direct image of an object; others use photometry to measure the amount of light coming from an object, and some telescopes use spectroscopy to measure the wavelength of light. They can either refract or reflect images and light. Optical telescopes are used to measure the light emitted by pulsars and supernovae.

Most optical telescopes are physically located in places where light pollution and water in the atmosphere are low and will not obstruct viewing. Consequently, many optical telescopes are located in high deserts or on mountain tops. The Keck Observatory in Hawaii, the Lick Observatory near San Jose, California, and the SETI Optical Telescope in Massachusetts (located at the Harvard Smithsonian Center for Astrophysics Oak Ridge Observatory) are some optical observatories used in the search for life on other worlds.

Overarching Questions

1. What is the likelihood of finding life in the universe?
2. What are the resources, technologies, and methods involved in using radio and optical astronomy for this search?
3. What progress and evolution has occurred in the field?
4. What resources are most important to success in the field?
5. What is the public interest in the topic?

Chairman SMITH. The Committee on Science, Space, and Technology will come to order.

And welcome to today's hearing "Astrobiology and the Search for Life in the Universe." A couple of preliminary announcements. One is that I want to thank C–SPAN for covering this hearing today. That shows the importance of the hearing in a lot of respects.

And I want to thank all the students from Herndon High School here as well. I understand you had a choice of hearings to attend, in fact you could attend almost any hearing you wanted to, and you chose this one because you thought it was the most interesting. And actually that is one of the purposes of today's hearing, and that is to inspire students today to be the scientists of tomorrow. And who knows? We may have some of those scientists in the audience right now who will be inspired by what they hear to study astrobiology or perhaps some of the other sciences as well. So we appreciate your attendance.

I will recognize myself for an opening statement and then the Ranking Member as well.

As we discover more planets around the stars in our own galaxy, it is natural to wonder if we may finally be on the brink of answering the question, "Are we alone in the universe?"

Finding other sentient life in the universe would be the most significant discovery in human history. Scientists estimate that there are 80 billion stars in the Milky Way galaxy. To date, more than 1,700 nearby planets have been found by the Kepler Space Telescope.

Last month, astronomers discovered the first Earth-like planet orbiting its star at a distance where liquid water could be present, a condition thought essential to life. Called Kepler-186f, it is only ten percent larger than the Earth and about 490 light years away.

The Transiting Exoplanet Survey Satellite, which will launch in 2017, and the James Webb Space Telescope, launching in 2018, will help scientists discover more planets with potential biosignatures.

The United States has pioneered the field of astrobiology and continues to lead the world in this type of research. A sample of professional papers published in Science magazine between 1995 and 2013 illustrates the significant growth and growing popularity of the field of astrobiology. Between 1995 and 2012, the number of papers published on astrobiology increased 10 times and the number of scientific reports that cited astrobiology increased 25 times.

Astrobiology is a serious subject studied by serious scientists around the world. Reflecting this interest, next September the Library of Congress and NASA will hold a 2-day astrobiology symposium on what the societal impacts could be of finding microbial, complex, or intelligent life in the universe.

Whether life exists on other planets in the universe continues to be a matter of debate among scientists. Around the world a number of astronomers listen to naturally occurring radio frequencies. They try to filter out the cosmic noise and interference of human-made satellites and spacecraft to find anomalies that could be signals from civilizations elsewhere in the universe.

The Allen Telescope Array at the SETI Institute, financed by Microsoft co-founder Paul Allen, and the Arecibo telescope in Puerto Rico are two well-known locations for conducting radio astronomy searches for life in the universe.

Recently, radio astronomers have detected pulsed signals that last only a few milliseconds. These "fast radio bursts" as they are called have caused scientists to speculate as to their cause. Some scientists hypothesize they could be from stars colliding or from an extraterrestrial intelligent source.

Other astronomers search for laser light pulses, instead of radio waves. Researchers at the SETI Optical Telescope, run by the Harvard Smithsonian Center for Astrophysics, the Columbus Optical SETI Observatory and the University of California at Berkeley, among others, use optical telescopes to try to detect nanosecond pulses or flashes of light distinct from pulsars or other naturally occurring phenomena.

I hope today's hearing will enable us to learn more about how research in astrobiology continues to expand this fascinating frontier. The unknown and unexplored areas of space spark human curiosity. Americans and others around the world look up at the stars and wonder if we are alone or is there life on other planets.

[The prepared statement of Mr. Smith follows:]

PREPARED STATEMENT OF CHAIRMAN LAMAR S. SMITH

As we discover more planets around the stars in our own galaxy, it is natural to wonder if we may finally be on the brink of answering the centuries' old question, "Are we alone in the universe?"

Finding other sentient life in the universe would be the most significant discovery in human history. Scientists estimate that there are 800 billion stars in the Milky Way. To date, more than 1,700 nearby planets have been found by the Kepler Space Telescope.

Last month, astronomers discovered the first Earth-like planet orbiting its star at a distance where liquid water could be present, a condition thought essential to life. Called Kepler 186f, it is only 10% larger than Earth and is 490 light years away.

The Transiting Exoplanet Survey Satellite, which will launch in 2017, and the James Webb Space Telescope, launching in 2018, will help scientists discover more planets with potential biosignatures.

The United States has pioneered the field of astrobiology and continues to lead the world in this type of research. A sample of professional papers published in Science magazine between 1995 and 2013 illustrates the significant growth and growing popularity of the field of astrobiology. Between 1995 and 2012, the number of papers published on astrobiology increased ten times and the number of scientific reports that cited astrobiology increased 25 times.

Astrobiology is a serious subject studied by serious scientists around the world. Reflecting this interest, next September the Library of Congress and NASA will hold a two day astrobiology symposium on what the societal impacts could be of finding microbial, complex or intelligent life in the universe.

Whether life exists on other planets in the universe continues to be a matter of debate among scientists. Around the world a number of astronomers listen to naturally occurring radio frequencies. They try to filter out the cosmic noise and interference of human-made satellites and spacecraft to find anomalies that could be signals from civilizations elsewhere in the universe.

The Allen Telescope Array at the SETI Institute, financed by Microsoft co-founder Paul Allen, and the Arecibo telescope in Puerto Rico are two well-known locations for conducting radio astronomy searches for life in the universe.

Recently radio astronomers have detected pulsed signals that last only a few milliseconds. These "Fast Radio Bursts" have caused scientists to speculate as to their cause. Some scientists hypothesize they could be from stars colliding or from an extraterrestrial intelligent source. Other astronomers search for laser light pulses, instead of radio waves. Researchers at the SETI Optical Telescope, run by the Harvard Smithsonian Center for Astrophysics, the Columbus Optical SETI Observatory and the University of California at Berkeley, among others, use optical telescopes to try to detect nanosecond pulses or flashes of light distinct from pulsars or other naturally occurring phenomena.

I hope today's hearing will enable us to learn more about how research in astrobiology continues to expand this fascinating frontier. The unknown and unexplored areas of space spark human curiosity. Americans and others around the world look up at the stars and wonder if we are alone or is there life on other planets.

Chairman SMITH. That concludes my opening statement, and the Ranking Member, the gentlewoman from Texas, Ms. Johnson, is recognized for hers.

Ms. JOHNSON. Thank you very much, Mr. Chairman, and good morning. In the interest of saving time I will forgo making an opening statement and instead I will simply want to welcome Dr. Shostak and Dr. Werthimer to this morning's hearing on the search for life, including intelligent life, in outer space. You both are distinguished researchers and I know that you will have thoughtful testimony to present, and this afternoon will determine whether we will have researchers to continue this.

So thank you and I yield back.

Chairman SMITH. Thank you, Ms. Johnson.

And I would like to introduce our witnesses at this point.

Our first witness, Dr. Seth Shostak, is a Senior Astronomer at the SETI Institute in Mountain View, California. He has held this position since 2001. Dr. Shostak has spent much of his career conducting radio astronomy research on galaxies. Dr. Shostak has written more than 400 published magazine and web articles on various topics in astronomy, technology, film, and television. He has also edited and contributed to nearly a dozen scientific and popular astronomy books. He has authored four books, including "Sharing the Universe: Perspectives on Extraterrestrial Life" and "Confessions of an Alien Hunter: a Scientist's Search for Extraterrestrial Intelligence." You can hear him each week as host of a one-hour-long radio program on astrobiology entitled "Big Picture Science."

Dr. Shostak received his bachelor's in physics from Princeton and his Ph.D. in astrophysics from the California Institute of Technology.

Our second witness, Dr. Dan Werthimer, has worked at the Space Sciences Laboratory at UC Berkeley since 1983. He is currently the Director of several of the lab's centers, including the SETI Research Center and the Center for Astronomy Signal Processing and Electronics Research.

Additionally, Mr. Werthimer serves as Chief Scientist for the lab's SETI@home program and Associate Director of their Berkeley Wireless Research Center. Mr. Werthimer co-authored "SETI 2020" and was the editor of "Bioastronomy: Molecules, Microbes, and Extraterrestrial Life" and "Astronomical and Biochemical Origins and the Search for Life in the Universe."

His research has been featured in many broadcast news stories such as on ABC and CBS and many major newspapers and magazines. His work also has reached a younger audience through Scholastic Weekly, a science magazine for kids.

Mr. Werthimer received his bachelor's and master's degrees in physics and astronomy from San Francisco State University.

I will recognize to start us off today Dr. Shostak and then we will go to Mr. Werthimer.

TESTIMONY OF DR. SETH SHOSTAK,
SENIOR ASTRONOMER AT THE SETI INSTITUTE

Dr. SHOSTAK. Thank you, Congressman Smith, for the opportunity to be here.

I am just going to give you a few big-picture thoughts on the search for life and in particular intelligent life, the kind of life that could uphold its side of the conversation as opposed to the microbial sort of life. This is obviously a subject of great interest to many people.

Let me just back up and say that when you read in the paper about the discovery of a new planet or something, water on Mars, you are looking at one of three horses in a race to be the first to find some extraterrestrial biology. The first horse is simply to find it nearby, and that is where the big money is. Rovers on Mars, the moons of the outer solar system. There are at least a half a dozen other worlds that might have life in our solar system. The chances of finding it I think are good, and if that happens, it will happen in the next 20 years, depending on the financing.

The second horse in that race is to build very large instruments that can sniff, if you will, the atmospheres of planets around other stars and maybe find oxygen in the atmosphere or methane, which, as you know, is produced by cows and pigs and things like that, but biology in any case. And—so you could find pigs in space, I suppose. That is again a project depending on funding that could yield results in the next two decades.

The third horse in that race is SETI, Search for Extraterrestrial Intelligence, and that idea, if you have seen the movie Contact you know what the idea is, is to eavesdrop on signals that are either deliberately or accidentally leaked off somebody else's world. That makes sense because in fact even we, only 100 years after Tesla and Marconi and the invention of practical radio, we already have the technology that would allow us to send bits of information across light years of distance to putative extraterrestrials.

Let me just tell you why I think they are out there, by the way. That—you know, it is unproven whether there is any life beyond Earth. That is the situation today. You have heard me say twice now that I think the situation is going to change within everyone's lifetime in this room. Okay. And the reason is we are—the universe is a fecund place for life. Congressman Smith has mentioned the number of stars in our galaxy. With respect, that number is actually rather larger. It is something like 200 to 400 billion stars, but we now know that at least 70 percent of them have planets. Recent results from NASA's Kepler telescope, an astoundingly successful instrument, suggest that one in five stars may have planets that are cousins of the Earth. What that means is that in our own galaxy there are tens of billions of other planets that are the kind you might want to build condos on and live. Okay. Tens of billions. And if that isn't adequate for your requirements, let me point out there are 150 billion other galaxies we can see with our telescopes, each with a similar complement of Earthlike worlds.

What that means is that the numbers are so astounding that if this is the only planet on which not only life but intelligent life has arisen, then we are extraordinarily exceptional. It is like buying trillions of lottery tickets and none of them is a winner. That would be very, very unusual. And although everybody likes to think that they are special, and I am sure you all are, maybe we are not that special. Certainly the history of astronomy shows that every time we thought we were special, we were wrong.

So what has been done so far, we have had various kinds of radio searches. I won't detail the technology. We have looked at much of the sky at fairly low sensitivity over a limited range of radio wavelengths, radio sections of the band. We have looked in particular directions at a few thousand star systems. In other words, we have just begun the search. The fact that we haven't found anything means nothing. It is like looking for megafauna in Africa and giving up after you have only examined one city block. And the reason the search has been so cramped and constricted so far is simply, to be honest, the fact that there is no funding for this. It is all privately funded. The total number of people in the world that do SETI for a living is fewer than the number of people in any row in the audience here behind me. That is the world total for this endeavor.

When are we going to find them? You have already heard me suggest that that may happen rather quickly. Let me just point out two other things. One, this is very interesting to the public because they have seen extraterrestrials on television and in the movies all their lives, okay. That also gives it a certain giggle factor. It is very easy to make fun of this. On the other hand, it would have been easy to make fun of Ferdinand Magellan's idea to sail around the Earth or Captain Cook to map the South Pacific. It is exploration. That is what this is.

The consequences are always, shall we say, salubrious. To find that there is life out there, intelligent life, would calibrate our position in the universe. It would, as Congressman Smith says, probably be the greatest discovery that humankind could ever make, and what is important is this is the first generation that has both the knowledge and the technology to do that.

[The prepared statement of Dr. Shostak follows:]

Written Text for Congressional Testimony

Using Radio in the Search for Extraterrestrial Intelligence

Seth Shostak, Senior Astronomer, SETI Institute
189 Bernardo Ave., #100, Mountain View, CA 94043, seth@seti.org

The question of whether we share the universe with other intelligent beings is of long standing. Written speculation on this subject stretches back to the classical Greeks, and it hardly seems unreasonable to suppose that even the earliest *Homo sapiens* gazed at the night sky and wondered if beings as clever as themselves dwelled in those vast and dark spaces.

What is different today is that we have both sufficient scientific knowledge and adequate communications technology to permit us to address this question in a meaningful way.

Finding extraterrestrial intelligence would calibrate humanity's place in the cosmos. It would also complete the so-called Copernican revolution. Beginning about 470 years ago, observation and scientific reasoning led to an accurate understanding of our place in the physical universe. The goal of SETI – the Search for Extraterrestrial Intelligence – is to learn our place in the intellectual universe. Are our cognitive abilities singular, or are they simply one instance among many?

Just as large sailing ships and the compass inaugurated the great age of terrestrial exploration at the end of the 15th century, so too does our modern technology – coupled to a far deeper understanding of the structure of the universe than we had even two decades ago – give us the possibility to discover sentient life elsewhere. SETI is exploration, and the consequences of exploration are often profoundly enlightening and ultimately of unanticipated utility. We know that our species is special, but is it unique? That is the question that SETI hopes to answer.

Why we think that life exists elsewhere

There is, as of now, no compelling evidence for biology beyond Earth. While the widely reported claims of fossilized microbes in a martian meteorite generated great excitement in 1996, the opinion of most members of the astrobiology community today is that the claims are unconvincing.

Nonetheless these same astrobiologists, if asked if they think it likely that extraterrestrial life is both commonplace and discoverable, would nod their heads affirmatively.

They would do so largely because of what we've learned in the past two decades concerning the prevalence of life-friendly cosmic habitats. Until 1995, we knew of no planets around other stars, habitable or otherwise. And yes, there was speculation that such worlds might be common, but that sunny thought was *only* speculation.

In the last two decades, astronomers have uncovered one so-called exoplanet after another. The current tally is approximately two thousand, and many more are in the offing thanks to continued analysis of data from NASA's enormously successful Kepler space telescope.

Estimates are that at least 70 percent of all stars are accompanied by planets, and since the latter can occur in systems rather than as individuals (think of our own solar system), the number of planets in the Milky Way galaxy is of order one trillion. It bears mentioning that the Milky Way is only one of 150 billion galaxies visible to our telescopes – and each of these will have its own complement of planets. This is plentitude beyond easy comprehension.

The Kepler mission's principal science objective has been to determine what fraction of this planetary harvest consists of worlds that could support life. The usual metric for whether a planet is habitable or not is to ascertain whether liquid water could exist on its surface. Most worlds will either be too cold, too hot, or of a type (like Jupiter) that may have no solid surface and be swaddled in noxious gases. Recent analyses of Kepler data suggest that as many as one star in five will have a habitable, Earth-size planet in orbit around it. This number could be too large by perhaps a factor of two or three, but even so it implies that the Milky Way is home to 10 to 80 billion cousins of Earth.

There is, in other words, more than adequate cosmic real estate for extraterrestrial life, including intelligent life.

A further datum established by recent research is that the chemical building blocks of life – the various carbon compounds (such as amino acids) that make up all terrestrial organisms – are naturally formed and in great abundance throughout the cosmos. The requisites for biology are everywhere, and while that doesn't guarantee that life will be spawned on all the worlds whose physical conditions are similar to our own, it does encourage the thought that it occurs frequently.

If even if only one in a thousand "earths" develop life, our home galaxy is still host to tens of millions of worlds encrusted by flora and fauna.

However, SETI is a class of experiments designed to find not just life, but technologically sophisticated life – beings whose level of intellect and development is at least equal to our own. So it is germane to ask, even assuming that there are many worlds with life, what fraction will eventually evolve a species with the cognitive talents of *Homo sapiens*? This is a question that's both controversial and difficult to answer.

As some evolutionary biologists (including most famously Ernst Mayr and Stephen Jay Gould) have pointed out, the road from early multicellular life forms (e.g., trilobites) to us is an uncertain one with many forks. For example, if the asteroid that wiped out the dinosaurs (and two-thirds of all other land-dwelling species) 65 million years ago had arrived in our neighborhood 15 minutes later, it could have missed the Earth. The stage might never have been cleared for the mammals to assert themselves and eventually produce us. This simple argument suggests that, while life could be commonplace, intelligence might be rare.

On the other hand, recent research has shown that many different species of animals have become considerably more clever in the last 50 million years. These include of course simians – but also dolphins, toothed whales, octopuses, and some birds. One plausible interpretation of these findings is that intelligence has so much survival value that – given a complex biota and enough time – it will eventually arise on any world.

We don't know what the truth is regarding the emergence of cognition. But finding another example of an intelligent species would tell us that *Homo sapiens* is not singular. The possibility of elucidating this evolutionary question is one of the most enticing motives for doing SETI experiments.

Finding extraterrestrial intelligence

Although encounters with intelligent aliens are a frequent staple of movies and television, the idea of establishing the existence of these putative beings by traveling to their home planets is one that will remain fiction for the foreseeable future. The planets that orbit the Sun may include other worlds with life (Mars, various moons of the planets Jupiter and Saturn). But they are surely devoid of any life that would be our cerebral equals. Intelligent beings -- assuming they exist – are on planets (or possibly large moons) orbiting other stars. Those are presently unreachable: Even our best rockets would take 100 thousand years to traverse the distance to the nearest other stellar systems. The idea that extraterrestrials have come here (the so-called UFO phenomenon), while given credence by approximately one-third of the populace, is not considered well established by the majority of scientists.

However, the methods used by SETI to discover the existence of intelligence elsewhere don't require that either we or they pay a visit. All we need do is find a signal, come to us at the speed of light. The first modern SETI experiment was conducted in 1960, when astronomer Frank Drake used an 85-foot diameter antenna at the newly constructed National Radio Astronomy Observatory in West Virginia in an attempt to "eavesdrop" on signals either deliberately or accidentally transmitted by beings light-years away. Drake used a very simple receiver, and examined two nearby star systems.

By contrast, later SETI experiments have made use of far more sensitive equipment, and have greatly expanded the scope of the search. Project Phoenix – a survey by the SETI Institute of 1,000 star systems – used antennas that ranged from 140 – 1,000 feet in diameter with receivers that could look for weak signals in ten million radio channels simultaneously. Today's efforts by the Institute use a small grouping of 42 antennas known as the Allen Telescope Array, situated in the Cascade Mountains of northern California. The advantage of this instrument is that it can be used for a very high percentage of time for SETI experiments, unlike previous campaigns that relied on antennas that were shared with radio astronomers doing conventional research projects. This latter circumstance greatly constrained the number of possible searches.

The other large radio SETI group in the U.S. is at the University of California, Berkeley. Their long-running Project SERENDIP uses the very large (1,000-foot diameter) antenna at Arecibo, Puerto Rico in a commensal mode. By piggybacking on this antenna, the Berkeley group gets virtually continuous use of the antenna, but the price is that they have no control of where it is aimed. However, over the course of several years, this random scrutiny covers roughly one-third of the sky. The receiver can simultaneously monitor more than 100 million channels, and some of the Berkeley data are made available for processing by individuals on their home computers using the popular screen saver, SETI@home. Approximately ten million people have downloaded the screen saver.

At the moment, the only other full-time radio SETI experiment is being conducted by a small group at the Medicina Observatory of the University of Bologna, in Italy.

Radio SETI searches preceded efforts to look for brief laser light pulses, known as optical SETI, largely because the development of practical radio occurred more than a half-century before the invention of the laser. Nonetheless, radio remains a favored technique for establishing the existence of intelligence beyond Earth. The amount of energy required to send a bit of information from one star system to another using radio is less than other schemes, and therefore it seems plausible that, no matter what other communication technologies intelligent species might develop, radio will always have a function. As a simple analogy: the wheel is an ancient technology for us, yet we use it every day and undoubtedly always will.

Radio SETI experiments have not yet detected a signal that is unambiguously extraterrestrial. Some people, both in and out of the science community, have ascribed undue significance to this fact, claiming that it indicates that no one is out there. While this may be comforting to those who would prefer to think that our species is the only one with the wit to comprehend the cosmos, it is a thoroughly unwarranted conclusion. Despite a many-decades long history of effort, our scrutiny of star systems still remains tentative. The number of star systems carefully examined over a wide range of the radio dial is no more than a few thousand. In the Milky Way, there are hundreds of billions of star systems. Consequently, our reconnaissance is akin to exploring Africa for megafauna, but one that has so far been limited to a square city block of territory.

While no one knows how prevalent signal generating civilizations might be, the more conservative estimates suggest that – to find a transmission that would prove others are out there – requires surveillance of a million star systems or more. This could be done in the near future, given the relentlessly increasing power of digital electronics. It is not hyperbolic to suggest that scientists could very well discover extraterrestrial intelligence within two decades' time or less, given resources to conduct the search.

However, funding for SETI is perennially problematic. The most ambitious SETI program, the one planned by NASA in the 1980s and 1990s, had scarcely begun observations when Congress canceled funding in the Fall of 1993. Since then, SETI efforts in this country have either been privately funded, or been an incidental part of university research. As a telling metric of the limitations of this approach, note that the total number of scientists and engineers doing full-time

SETI in this country is approximately one dozen, or comparable to the tally of employees at a car wash.

Progress and evolution of radio SETI

A rough and ready estimate suggests that today's radio SETI experiments are about 100 trillion times more effective – as judged by speed, sensitivity, and range of radio frequencies investigated – than Frank Drake's pioneering 1960 search. The rapid development of both analog and digital electronics has spinoffs that are accelerating the capabilities of SETI.

As example, in 1980 typical SETI efforts sported receivers able to monitor 10 thousand channels simultaneously. Today's experiments sport 10 – 100 million channels, causing a thousand-fold increase in search speed.

Speed is essential to success. As mentioned above, conservative estimates of the prevalence of broadcasting societies hint that – in order to find a signal from another species – our SETI experiments will need to "listen" in the direction of at least 1 million stellar systems. Cheaper digital technology, which can be read as greater compute power, immediately leads to receivers with more channels – which means that it takes less time to check out all the interesting frequencies for a given SETI target.

In the case of antenna arrays, cheaper computing can also speed observations by increasing the number of star systems looked at simultaneously. As example, the Allen Telescope Array currently has the ability to examine three such systems at once. But this could be increased to hundreds or even thousands with more computing power – bringing with it a concomitant augmentation of speed.

Current and future resources

As noted above, the level of radio SETI effort today is small, employing roughly a dozen full-time scientists and engineers. At the height of the NASA SETI program (1992), the annual budget for this activity was $10 million, or one-thousandth of the space agency's budget. This supported equipment development and observations for a two-pronged strategy – a low-sensitivity survey of the entire sky, and a high-sensitivity targeted search of the nearest thousand star systems. The number of scientists involved was five times greater than today.

The financial support for all radio SETI efforts in the United States now is approximately 20 percent of the earlier NASA program, and comes from either private donations or from research activities at the University of California. This is, frankly, a level inadequate for keeping this science alive. The cost of developing and maintaining the requisite equipment and software, as well as paying for the scientists and engineers who do the experiments, is – at minimum – $5 million annually. Without this level of funding, the U.S. SETI efforts are likely to be overtaken by Asian and European initiatives (such as the Square Kilometer Array) in the next decade.

ETI is exploration. There's no way to guarantee that if only sufficient effort is made, success will inevitably follow. Like all exploration, we don't know what we'll find, and it's possible that we'll not find anything. But if we don't search, the chances are good that the discovery of intelligence elsewhere in the cosmos will be made by others. That discovery will rank among the most profound in the history of humankind. The first evidence that we share the universe with other intelligence will be viewed by our descendants as an inflection point in history, and a transformative event.

The public's interest

The idea of extraterrestrials resonates with the public in a way that little of the arcane research of modern science does. While much was made of the discovery of the Higgs boson in 2012, people who weren't schooled in advanced physics had a difficult time understanding just why this was important, and what justified the multi-billion dollar price tag of the collider used in its discovery.

The idea of life in space on the other hand is science that everyone grasps. Countless creatures from the skies infest both movies and television. In addition, the techniques of SETI – while complex in detail – are simple in principle. Carl Sagan's novel and movie, "Contact", enjoyed considerable popularity, and familiarized millions with the technique of using radio to search for extraterrestrials. Documentaries on SETI and the search for life in general can be found on cable television every week. Compare that with the frequency of programming on, say, organic chemistry.

In other words, SETI is an endeavor that everyone "gets". And that includes school kids. This makes the subject an ideal hook for interesting young people in science. They come for the aliens, but along the way they learn astronomy, biology, and planetary science. Even if SETI fails to find a signal for decades, it does great good by enticing youth to develop skills in science.

It's even possible that we are hard-wired to be interested in extraterrestrial life, in the same way that we are programmed to be interested in the behavior of predators. The latter has obvious survival value (and might explain why so many young people are intrigued by dinosaurs!) Our interest in "aliens" could simply derive from the survival value of learning about our peers. Extraterrestrials are the unknown tribe over the hill – potential competitors or mates, but in any case someone we would like to know more about.

There's no doubt that SETI occasionally provokes derision. It's easy to make fun of an effort whose goal is to find "space aliens." But this is to conflate science fiction with science. As our telescopes continue to peel back the film that has darkened our view of the cosmos since *Homo sapiens* first walked the savannahs, we are learning that the Earth is only one of 100,000 billion billion planets, spinning quietly in the vast tracts of space. It would be a cramped mind indeed that didn't wonder who might be out there.

Biographical Sketch: Seth Shostak

Seth is the Senior Astronomer at the SETI Institute, in Mountain View, California. He has an undergraduate degree in physics from Princeton University, and a doctorate in astronomy from the California Institute of Technology. For much of his career, Seth conducted radio astronomy research on galaxies, publishing approximately sixty papers in professional journals. For more than a decade, he worked at the Kapteyn Astronomical Institute, in Groningen, The Netherlands, using the Westerbork Radio Synthesis Telescope. He also founded and ran a computer animation company.

Seth has written more than four hundred published magazine and web articles on various topics in astronomy, technology, film and television. He lectures on astronomy and other subjects at various academic venues, and gives approximately 60 talks annually at both educational and corporate institutions. Seth has been a Distinquished Speaker for the American Institute of Aeronautics and Astronautics. He also Chaired the International Academy of Astronautics' SETI Permanent Committee for a decade.

Frequently interviewed for radio and TV, Seth is the host of a one-hour weekly radio program on astrobiology entitled "Big Picture Science"

Seth has edited and contributed to nearly a dozen books. His first popular tome, "Sharing the Universe: Perspectives on Extraterrestrial Life" appeared in March, 1998, followed by "Cosmic Company" in 2002. He has also co-authored an astrobiology text, "Life in the Universe" that is now in its third edition, and his latest trade book is "Confessions of an Alien Hunter". In 2004, he won the Klumpke-Roberts Prize for the popularization of astronomy.

Chairman SMITH. Thank you, Dr. Shostak.
And, Mr. Werthimer.

TESTIMONY OF MR. DAN WERTHIMER, DIRECTOR OF THE SETI RESEARCH CENTER AT THE UNIVERSITY OF CALIFORNIA, BERKELEY

Mr. WERTHIMER. Thanks for the opportunity—thank you for the opportunity to talk to you about this question, are we alone? Is anybody out there?

Can you guys show the slides? I want to walk you through some of the SETI experiments that we and other people are doing.

Mr. WERTHIMER. So, as Seth mentioned, this NASA Kepler mission, from that we have learned that there are a trillion planets in our Milky Way galaxy. That is more planets than there are stars, lots of places for life. And we have learned that a lot of these planets are what we call Goldilocks planets, at the right distance where it is not too hot, not too cold, rocky planets, some have liquid water. So there could be a lot of life out there.

So how are we getting in touch? Well, one of the ideas is that earthlings have been sending off radio, television, radar signals out into space for the last 75 years. The early television shows like I Love Lucy, Ed Sullivan have gone past 10,000 stars. The nearby stars have seen the Simpsons. So you could turn that around. If we are broadcasting, maybe other civilizations are sending signals in our direction either leaking signals the way that we unintentionally send off signals or maybe a deliberate signal.

They could be sending laser signals, and there are a number of projects looking for laser signals. This is a project that Harvard University, a very clever project, this is a project at Lick Observatory. There is also a project at the—in Hawaii at the Keck Telescope looking for laser signals.

People are also looking for radio signals. Our group uses the world's largest radio antenna. We call it a radio telescope. This is the Arecibo telescope in Puerto Rico. It is 1,000 feet in diameter. It holds 10 billion bowls of cornflakes. We haven't actually tried that. It is operated by the National Science Foundation, and most astronomers would be lucky to use this telescope a day or two a year. We figured out a way to use the telescope at the same time that other scientists are using it so we can actually collect data all year round, all day. We are collecting data right now as we talk to you.

Now, that is actually a problem. So even though we have got the world's largest telescope all year round, it creates an enormous amount of data. And to analyze the data we asked volunteers for help. They—if you—you can help us by running a program on your home computer or your laptop or your desktop computer. You install a program called SETI@home. It is a screensaver program, and the way—we take the data from the world's largest telescope and we break it up into little pieces. Everybody gets a different piece of the sky to analyze. Then you install this program and it pops up when you go out for a cup of coffee and the computer goes through the data looking through all the different frequencies and signal types. This is what it looks like when it is running on your computer at home.

It takes a few days to analyze that data looking for interesting signals. And then when it finds interesting signals, it sends them back to Berkeley and then you get a new chunk of data, different part of the sky to work on.

If you are the lucky one that finds that faint murmur from a distant civilization, you might get the Nobel Prize, but there is a catch. The Nobel Prize—you have to maybe share with a lot of people. There are millions of people that have downloaded the SETI@home screensaver. They are split out into 200 countries. It is—together, the volunteers have formed one of the most powerful supercomputers on the planet and they have enabled the most sensitive search for extraterrestrial signals that anybody has ever done. So we are very grateful to the volunteers.

And now we have made that more general so that you can participate in not just SETI with your home computer but you can participate in lots of projects. There is climate prediction projects, there is a gravity wave project, there is protein folding. You can look for malaria drugs, HIV drugs, cancer drugs and you can allocate how you want your spare computing cycles to be used on your home computer.

One of the new projects we are working on is called Panchromatic SETI, and we are asking universities and observatories around the world to look at a lot of different wavelength bands, a lot of different frequencies. We are targeting the very nearest stars and we are trying to cover all the different bands that come through the Earth's atmosphere. We are looking at radio frequencies, we are looking at infrared frequencies or wavelengths, and we are looking at also optical frequencies looking for laser signals. And this will be an extremely comprehensive search because we have got eight different telescopes that we are using and looking at all these different bands but only targeting the nearby stars.

Another project that we are just launching this year is called Interplanetary Eavesdropping, and the idea of this project is that there may be signals going back and forth between two planets in a distant solar system. For instance, maybe eventually we will have machines or people on Mars and we will have radio communication or laser communication between our two planets. Well, put it the other way. A distant civilization may have colonized a planet in their own solar system and there may be radio or laser signals going back and forth between those two planets. And now with the Kepler spacecraft we know exactly when two planets in a distant solar system are lined up with Earth so we can schedule our observations and target that and see if we can intercept those signals going back and forth between two distant planets. We are using that—the Green Bank telescope in West Virginia to do that experiment.

While we haven't found ETs so far but we have made a lot of interesting discoveries. We have discovered a planet made out of solid diamond. We have made the first maps of the black hole of the center of the galaxy. These instruments are used in all kinds of things, in brain research which may eventually control prosthetic arms, but we haven't found ET so far. We are still working on it. We are just getting in the game. We have only had radio 100 years. We are just learning how to do it. It is like looking for a needle in a haystack but I am optimistic in the long run.

The reason I am optimistic in the long run is that the SETI is limited by computing technology, which is growing exponentially. It is limited by telescope technology. China is building a huge telescope, bigger than Arecibo. The Australians and South Africans and the Europeans are working on a huge telescope made out of thousands of dishes combined to make a giant telescope.

And I think I will stop there. I have got a couple of poems that I could read you from the volunteers but I am out of time. Thank you very much.

[The prepared statement of Mr. Werthimer follows:]

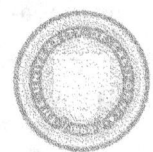

STATEMENT OF DAN WERTHIMER
DIRECTOR, BERKELEY SETI RESEARCH CENTER
UNIVERSITY OF CALIFORNIA, BERKELEY

TO THE

HOUSE COMMITTEE ON SCIENCE, SPACE, AND TECHNOLOGY
UNITED STATES HOUSE OF REPRESENTATIVES
MAY 21, 2014

Mr. Chairman and members of the committee:
Thank you for the invitation to discuss SETI, the Search for Extraterrestrial Intelligence.

SETI experiments are trying to determine whether other intelligent, technologically capable, life exists in the universe, to answer the question "Are we alone?" or "Is anybody out there?"

In the last fifty years, evidence has steadily mounted that the components and conditions we believe necessary for life are common and perhaps ubiquitous in our galaxy. Planets beyond our solar system, while once relegated to the domain of speculation, are now known to be common and numerous. Nevertheless, no evidence exists for the presence of life outside of the Earth. However, on our own planet, life is known to have arisen early and flourished. And while the propensity for evolution of intelligence from basic forms of life is not currently well understood, it appears that intelligence has imparted a strong evolutionary advantage to our own species. The possibility that life has arisen elsewhere, and perhaps evolved intelligence, is plausible and warrants scientific inquiry.

From NASA's Kepler mission we've learned there are roughly one trillion planets in our Milky Way galaxy; three times more planets than stars. Billions of these planets are Earth sized and in the "habitable" or so called "Goldilocks" zone - not too distant from their host star (too cold), and not too close to their star (too hot). And there are billions of other galaxies outside our Milky Way galaxy - plenty of places where life could emerge and evolve.

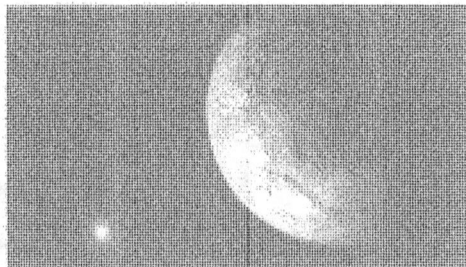

Figure 1: Artist's conception of Kepler 186f, an earth sized extrasolar planet in the habitable zone.

There may even be primitive extraterrestrial life in our own solar system, perhaps on a moon of Jupiter or Saturn. Europa, one of Jupiter's moons, is thought to have a liquid water ocean beneath its icy surface, perhaps a good environment for life as we know it.

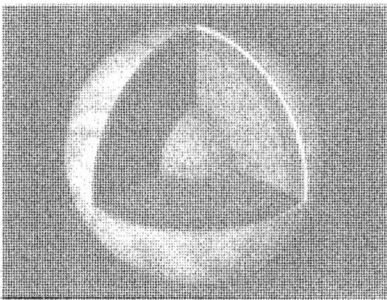

Figure 2: Cut away view of Europa, one of Jupiters moon, which has a liquid water ocean (blue) under a crust of ice (white), suitable for carbon based life. (artist depiction)

The universe is likely to be teeming with primitive life. Our growing knowledge of both terrestrial biology and extraterrestrial environments are steadily reinforcing the notion that there is nothing particularly unusual about the Earth and our Solar System. If the same processes that led to life's emergence on our own planet are at work on other worlds there is no barrier to life emerging there as well. Indeed many scientists are actively working along this line of reasoning. It may be only a few years before we have detected evidence of life beyond the Earth in spectrographic analyses of extrasolar planet atmospheres or in one of the myriad of current or proposed in-situ sampling missions exploring our Solar System. However, the deeper question of the processes of life's evolution that might lead to intelligence is addressable only by much more select means. The creation of technology, and especially environmental modification by that technology, is the only known tracer of intelligence detectable over interstellar distances. Radio communication in particular is a superb probe of extraterrestrial technology and is in fact the most detectable signature of our own technology.

SETI programs are not searching for primitive life; instead, SETI programs use the world's largest radio and optical telescopes to search for evidence of advanced civilizations and their technology on distant extrasolar planets.

Earth's civilization has been sending radio and television signals into space for roughly 85 years. Travelling at the speed of light, early television shows like I Love Lucy and the Ed Sullivan show have gone past tens of thousands of stars. Nearby stars have "seen" the Simpsons. Humans use powerful radars to monitor nearby space, and utilize bright lasers as adaptive optics "guide stars" for terrestrial optical observatories. Perhaps other civilizations, if they are out there, emit radio signals, navigational beacons, laser beams, or other signals that Earthlings could detect. Such signals could be accidental - an artifact of their technology, much the way that earth's television and radar leaks into space, or perhaps extraterrestrial civilizations might transmit deliberate signals for the purpose of interstellar communication.

When will Earthlings discover other civilizations? SETI experiments are in their early stages. We've only had radio for 100 years in the four billion year history of life on our planet. We are just now developing the tools and technologies that might detect distant civilizations. There could be radio or laser signals from extraterrestrial civilizations reaching our planet right now, but we would most likely not detect these signals with these early SETI projects. SETI is limited by computing and telescope technologies, and both of these technologies are growing rapidly. I'm optimistic in the long run, if there are signals from other civilizations, Earthlings will eventually be capable of detecting them.

SETI Observing Programs: There are about two dozen scientists on our planet who conduct SETI observations. Two thirds of these researchers are in the USA.

Professor Paul Horowitz leads an "optical SETI" program at Harvard, scanning the sky for laser signals. The possibility of detecting optical and infrared emission from extraterrestrial lasers was first suggested in 1961, well before the widespread proliferation of human laser technology in the later 20th century. In the last two decades, both pulsed and spectroscopic optical SETI searches have been conducted. The field of pulsed optical SETI rests on the observation that humanity could build a pulsed optical transmitter (using, for example, a National Ignition Facility-like laser coupled to a large optical telescope) that would easily be detectable at interstellar distances. When detected, the nanosecond-long pulses would be a factor of about 1000 brighter than the host star of the transmitter during their brief flashes. Such nanosecond-scale optical pulses are not known to occur naturally from any astronomical source.

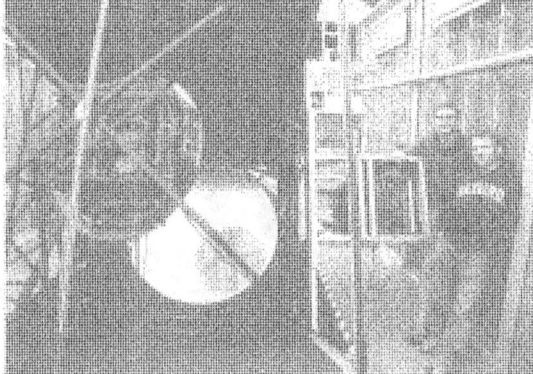

Figure 3: Harvard's Optical SETI observatory scans the sky for laser signals from distant civilizations. The 72 inch diameter telescope is shown with graduate students Curtis Mead (right), and Andrew Howard (left), now an extrasolar-planet-hunting Professor of Astronomy at the University of Hawaii. Harvard is using their innovative telescope and instrumentation to conduct the first all sky survey for laser signals.

The SETI Institute has led a number of SETI programs, and is currently conducting a search for radio signals using the Allen Telescope Array in Northern California. My colleague, Seth Shostak, from the SETI Institute, will hopefully review this project in his testimony today.

Figure 4: The Allen Telescope Array, in Northern California, is used by the SETI Institute to search for radio signals from extraterrestrial civilizations.

The Berkeley SETI Research Center at the University of California and our collaborators are conducting a variety of radio, infrared, and optical SETI searches using nine different observatories, including the National Astronomy and Ionosphere Center's Arecibo telescope in Puerto Rico, the National Radio Astronomy Observatory's Robert C. Byrd Green Bank telescope in West Virginia, the Keck telescope in Hawaii, several telescopes at Lick Observatory in California, the Infrared Spatial Interferometer in California, the Combined Array for Research in Millimeter-wave Astronomy in California, and the Low Frequency Array in Europe.

Berkeley's most well-known SETI project is SETI@home. SETI@home uses several telescopes to survey the sky, but most of the data comes from the world's largest single dish radio telescope, the 1000 foot diameter Arecibo telescope. The Arecibo Observatory is the most sensitive radio telescope on the planet, and as such represents our best chance of detecting the faintest radio whispers from advanced civilizations. Most scientists would be lucky to get a day or two each year to conduct observations using the Arecibo telescope, but our group pioneered a technique to use the telescope at the same time that other scientists are using the telescope for their own research. Using this "piggyback" or "commensal" SETI strategy, we are able to observe almost all year round on the world's largest telescope.

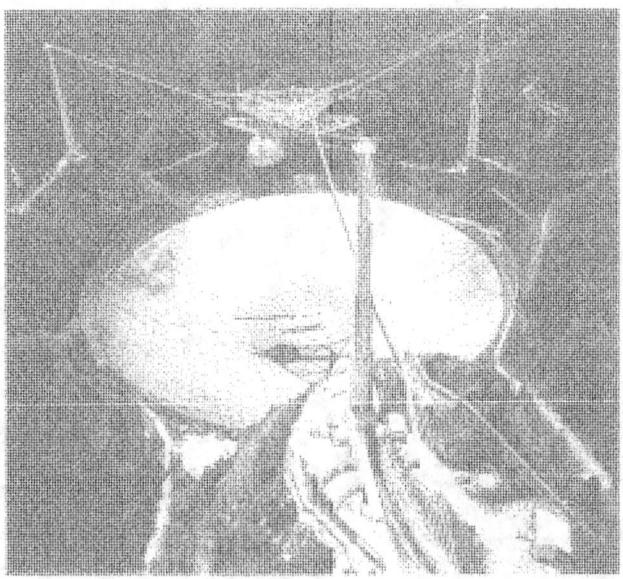

Figure 5: The National Astronomy and Ionosphere Center's
1000 foot diameter telescope in Arecibo, Puerto Rico.

To analyze the incoming data from the Arecibo telescope requires an enormous amount of computing power, which is needed to search through the many different places on the sky, frequencies (or "channels") and signal types (pulses, narrow and wide bandwidth signals, drifting signals…) that another civilization might be broadcasting. To analyze the hundreds of terabytes of data we collect, we ask volunteers around the world for help. Utilizing their laptop and desktop computers at home, office or school, volunteers can download the SETI@home screensaver program. When a volunteer's computer isn't being used, the screensaver program automatically downloads a small chunk of our data and goes to work searching for a rich variety of signal types. When the analysis is completed, typically after a few days, the program sends back the results of the analysis to our database, and the screen saver program gets a new chunk of data to work on. Millions of volunteers, in 226 countries, have downloaded the SETI@home screensaver. The SETI@home volunteers have formed one our planet's most powerful supercomputers and have enabled the world's most sensitive SETI search. SETI@home project scientist Dr. Eric Korpela is currently testing new SETI@home software that can run on cell phones, so that millions more will be able to participate in a global science project, allowing us to conduct our most thorough search ever.

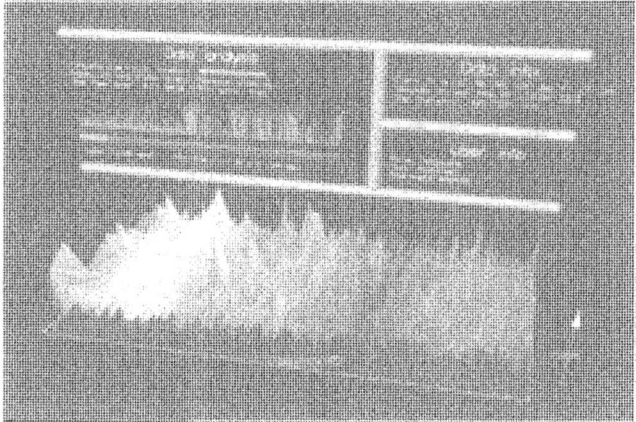

Figure 6: The SETI@home Screen Saver Program

One of the new SETI programs our research group is launching this year is called "panchromatic SETI". Led by Dr. Andrew Siemion, the project will search very nearby stars as well as stars determined to be most likely to host a solar-like exoplanet system. The experiment will use six different telescopes equipped with powerful spectrometers capable of examining five billion radio channels simultaneously and specialized infrared and visible light detectors, spanning the electromagnetic spectrum from the lowest frequencies that can pass through our planet's ionosphere, through infrared, and up to optical light – from 50 MHz to 500 THz. This experiment is targeting a wide range of possible signal types indicative of the presence of advanced technology. These observations will be the most sensitive and comprehensive SETI searches of these stars ever performed.

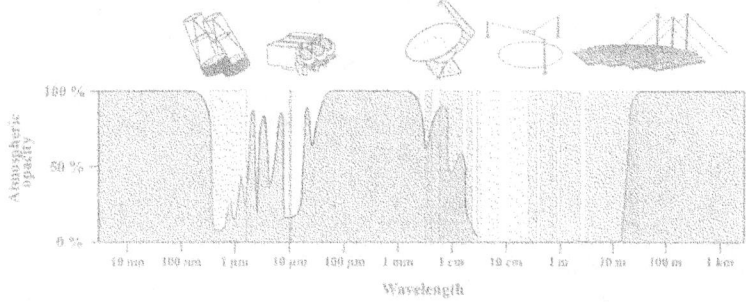

Figure 7: Wavelengths covered in the new "Panchromatic SETI Project" observing nearby stars, plotted against "atmospheric opacity," a measure of the amount of light that can get through the Earth's atmosphere. Several optical, infrared, and radio telescopes will be used to cover many of the wavelength bands that get through the Earth's atmosphere.

Another new project we are working on this year is dubbed "eavesdropping SETI." Someday we will colonize Mars – if so, we will want to communicate with people and machines on Mars or other planets in our solar system. Similarly, perhaps other civilizations are sending radio or laser signals between planets within their own solar system. Using recent data from NASA's Kepler mission, we can predict exactly when two extrasolar planets will be lined up with Earth, and we can schedule our observations to attempt to detect another civilization's inter-planetary communication.

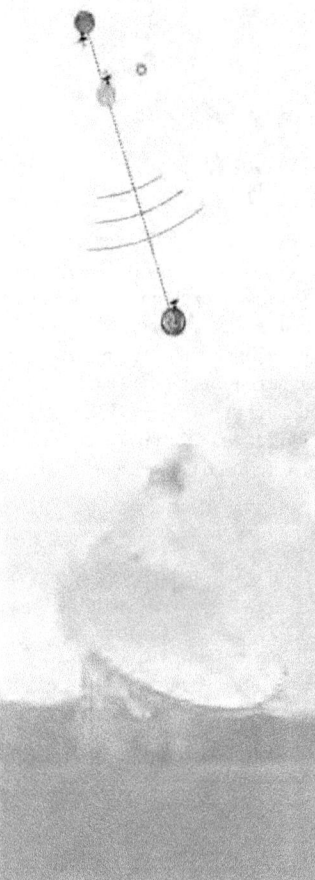

Figure 8: An illustration of Berkeley's "eavesdropping SETI" project targeting a conjunction of two planets in an exoplanetary system along a common line of sight. The paradigm envisioned here is that one or more planets in a system may be involved in communication or remote-sensing within their orbital plane.

Figure 9: The National Radio Astronomy Observatory's Robert C. Byrd telescope in Green Bank, West Virginia, utilized by the University of California Berkeley's eavesdropping and panchromatic SETI projects. The 100 meter diameter telescope is the largest fully steerable radio telescope on the planet, and with observing capabilities extending up to 100 GHz.

Spin Offs from SETI

Education and Outreach: SETI researchers haven't found ET yet, but there have been many spin offs and interesting discoveries along the way. SETI@home has engaged millions of volunteers in a global science project. Thousands of students are running SETI@home as part of their science curriculum. The question "Are we alone?" touches on many disciplines, including physics, astronomy, chemistry, biology, engineering, and computer science.

Volunteer computing and citizen science: Led by Dr. David Anderson, our SETI group developed general purpose open source software for public participation supercomputing. Using this software, millions of volunteers are participating in dozens of scientific computing projects, including malaria, cancer, and HIV drug research, climate modeling, pulsar searching, protein folding and SETI. Participants can use their spare computing cycles for the science projects they are most interested in.

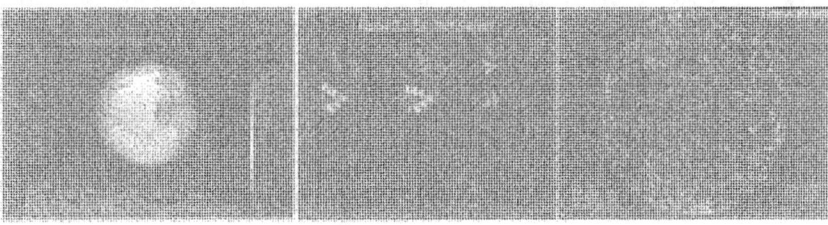

Figure 10: Three of the many volunteer computing projects that use software derived from SETI@home.

SETI Instrumentation Spin Offs: The powerful signal processing instrumentation that we originally developed for SETI is now widely used by radio astronomy observatories. These SETI instruments were adapted and used to produce the first images of the black hole at the center of our galaxy. The SETI instrumentation was also adapted and used to discover a planet made of solid diamond, as well as discover the most massive neutron star ever found, leading to an improved understanding of high density matter. The instruments have even been used in brain research, which may eventually lead to the control of prosthetic limbs.

Self-Driving Cars: SETI has trained students in electrical engineering, computer science, physics, and astronomy. One of our engineering graduate students, Pierre Droz, started a company that developed self-driving cars and tractors. The company delivered a pizza without a driver. Google purchased Droz's company.

Fragility of SETI Research

Despite widespread public support, there's not much funding for SETI research, and the funding fluctuates wildly. Our SETI program at the University of California currently receives roughly a million dollars a year in research grants, from NASA, the National Science Foundation, the Templeton Foundation, and other private donors.

Very few scientists and institutions work on SETI projects – there are about 24 scientists on the planet that carry out SETI observations. We are working to train new researchers, and help other universities and observatories launch new SETI projects.

The two best radio telescopes on the planet for SETI are in a budget crisis. The NSF is planning to discontinue funding for the Robert C. Byrd Green Bank telescope in West Virginia, and NSF astronomy has had to drastically cut support of the Arecibo telescope as well. Luckily, Arecibo is used for both atmospheric and asteroid research (Arecibo's radar can measure asteroid orbits extremely precisely, and can predict if Earth is in danger of being hit by an asteroid), so NASA and other NSF divisions have been able to step up their support and keep the Arecibo telescope running so far. However, the budget is extremely tight, and both telescopes are in jeopardy. Meanwhile, China is building an extremely large radio telescope, 500 meters in diameter (larger than Arecibo), and the multi-billion dollar International Square Kilometer Array Telescope project is well underway. These may soon become the world's preeminent radio SETI observatories, but the United States is not involved in either of them.

SETI Poetry

SETI@home volunteers have helped SETI in many ways, by building one of our planet's largest supercomputers and helping develop the software. Some have donated funds to keep the project running and growing. Some have composed music and literature, and thousands have written haikus. Here are two haikus composed by SETI@home volunteers:

Searching for ET
Answers are revealed
About ourselves   (Paula Cook)

One Million Earthlings
Bounded by Optimism
Leave Their PC's On   (Dan Seidner)

Are We Alone?

The Search for Extraterrestrial Civilizations
with help from Eight Million Volunteers

Dan Werthimer

Director, Berkeley SETI Research Center
University of California, Berkeley

SETI@home

NASA's Kepler Mission

- Determine the frequency of Earth-size and larger planets in the habitable zone of sun-like stars

- Determine the size and orbital period distributions of planets

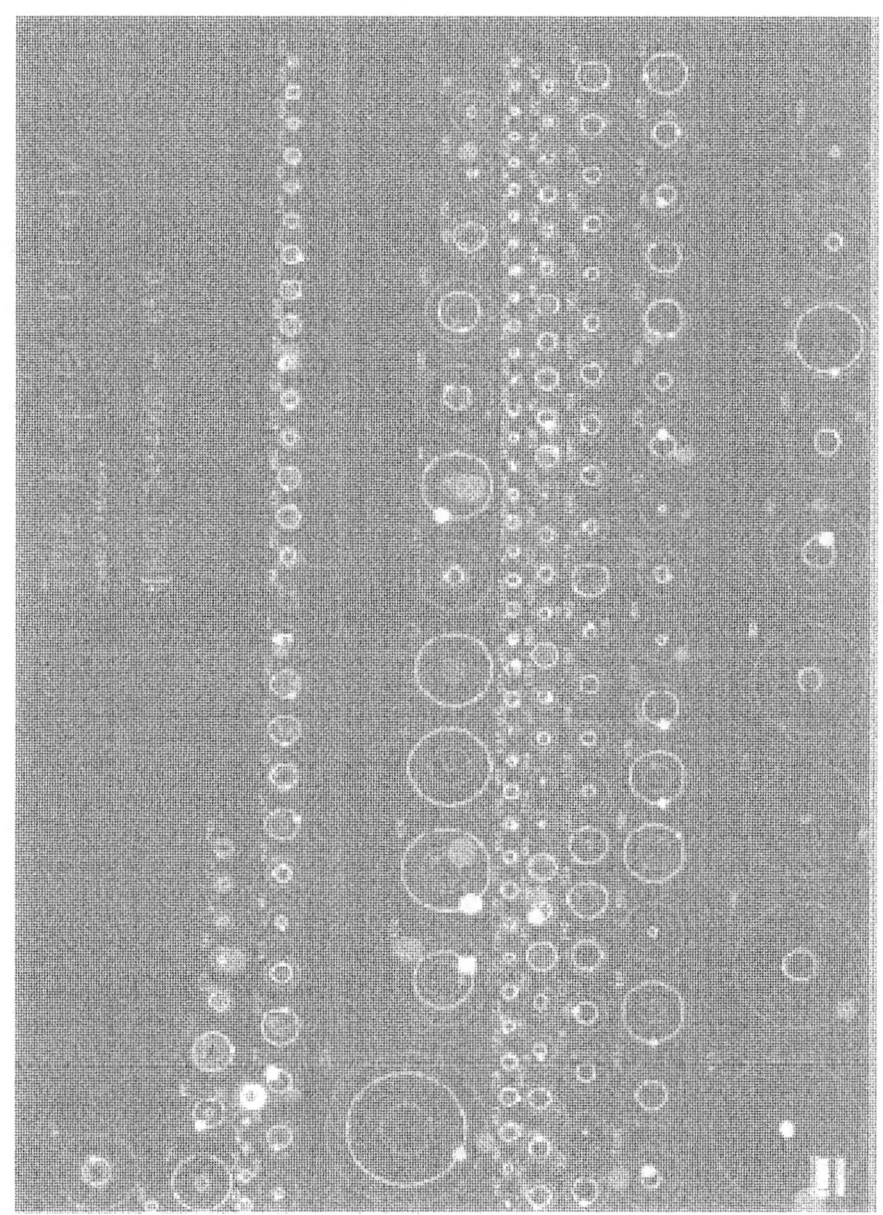

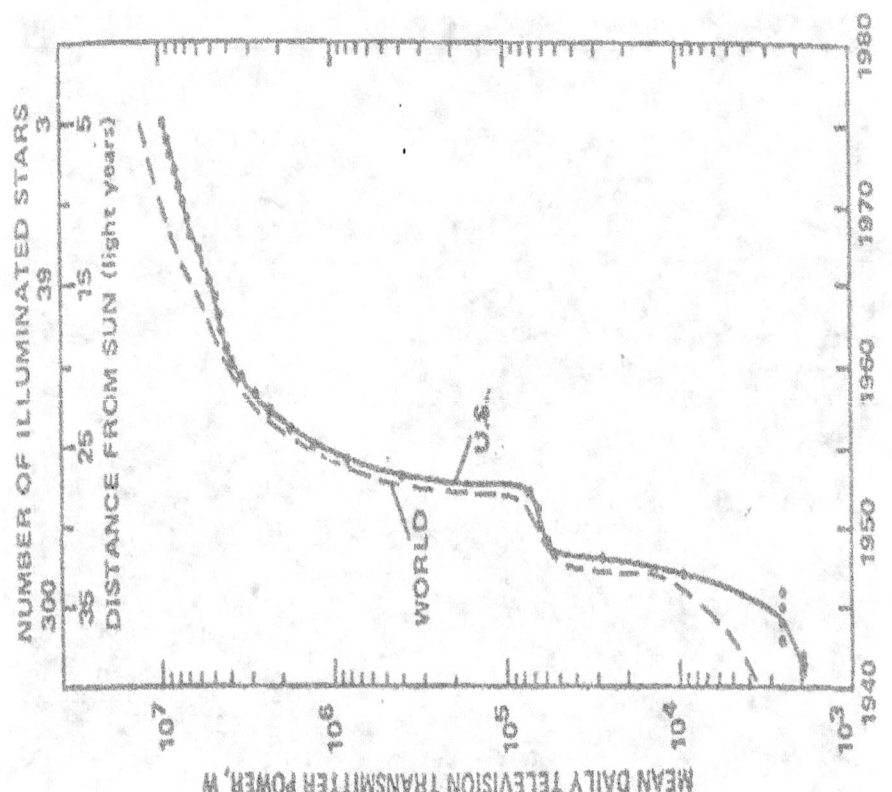

33

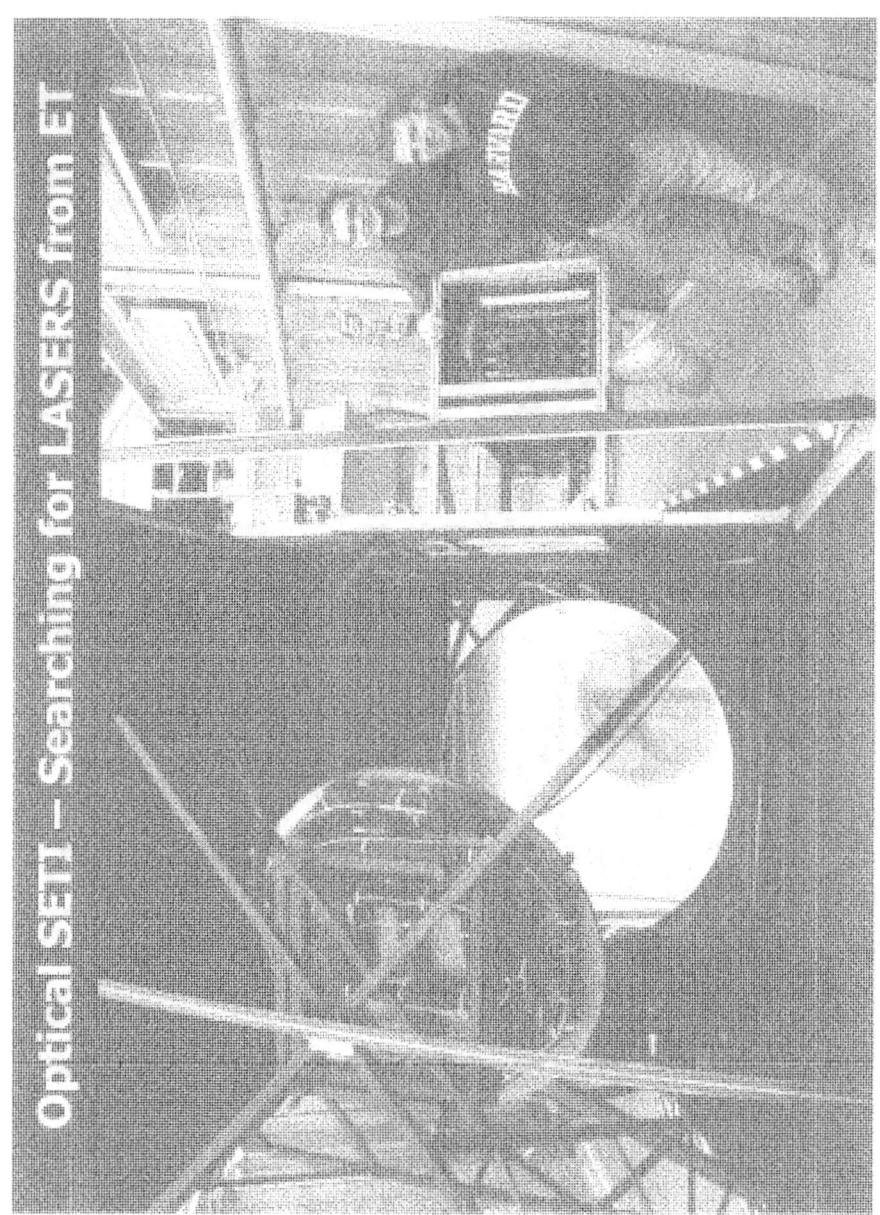

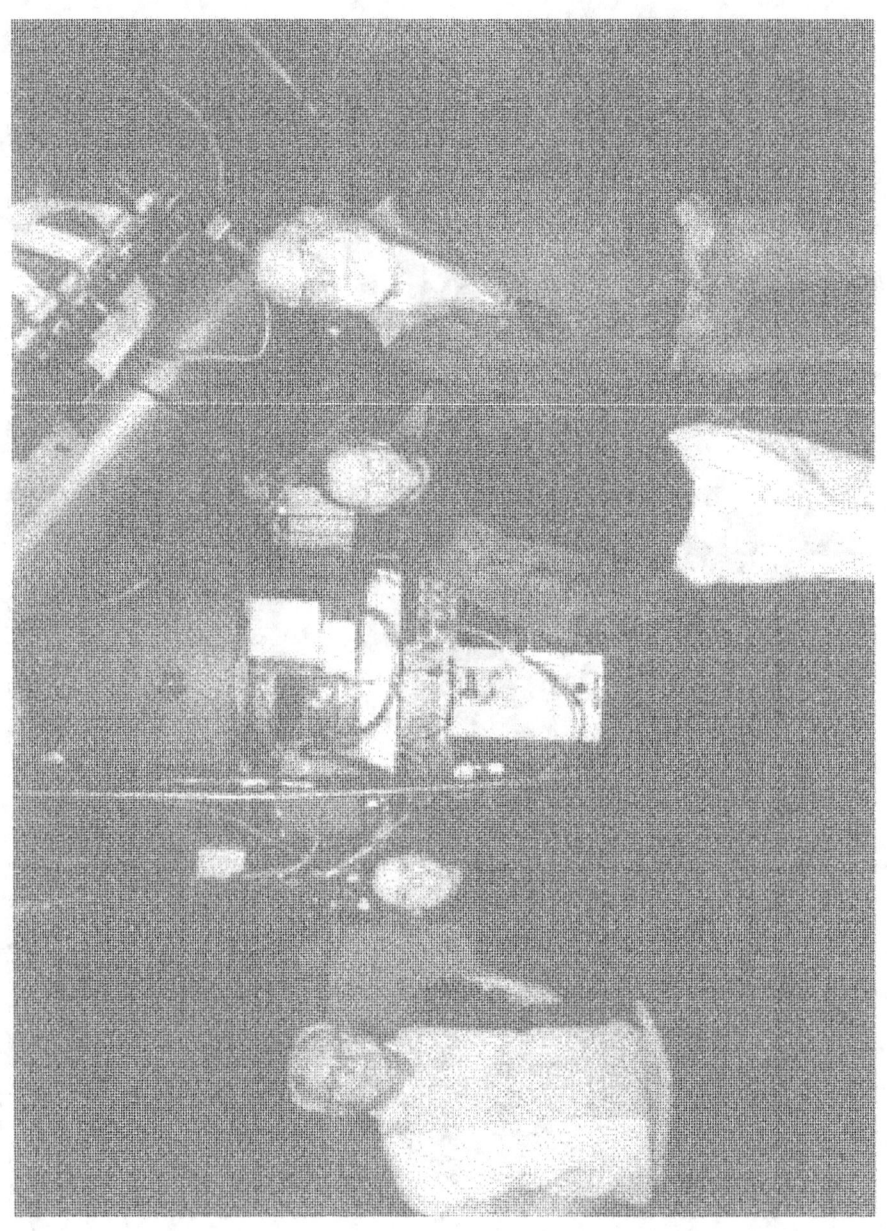

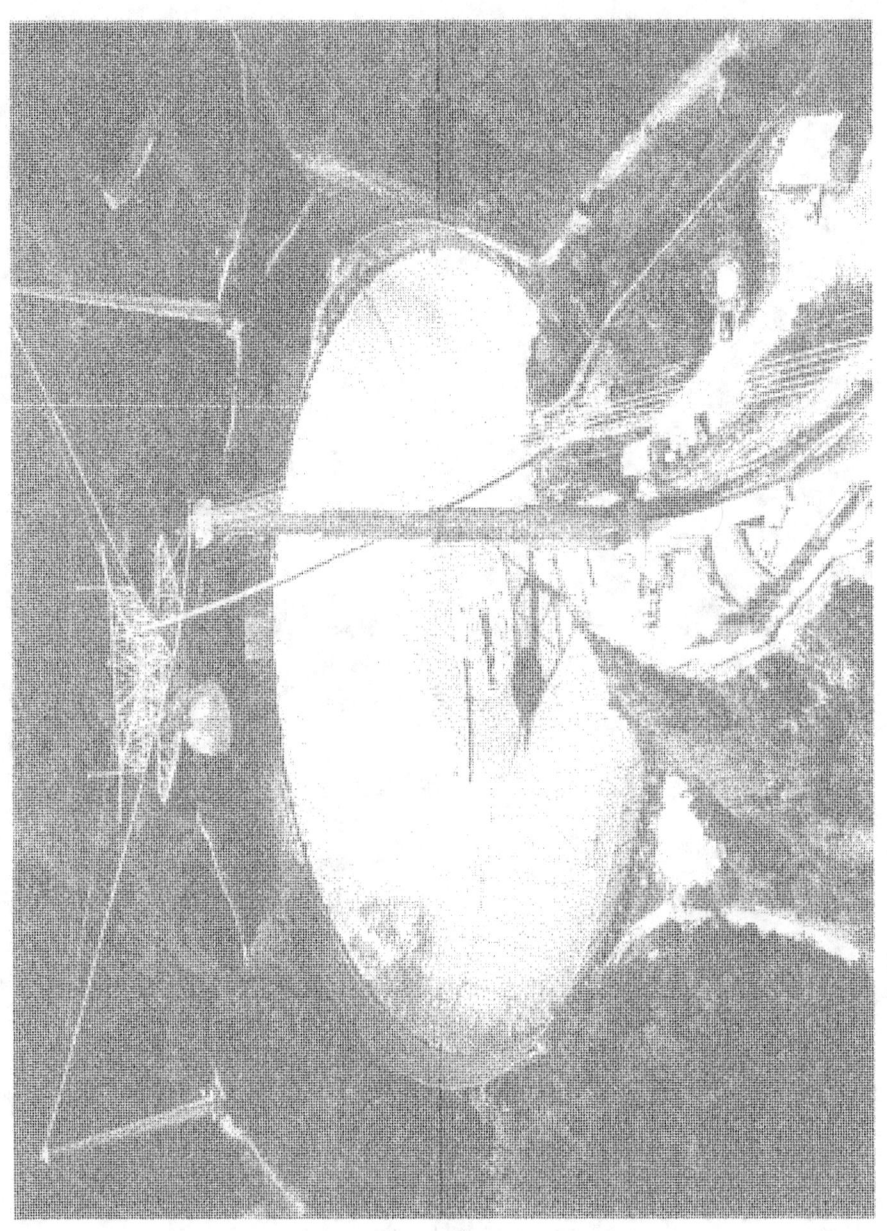

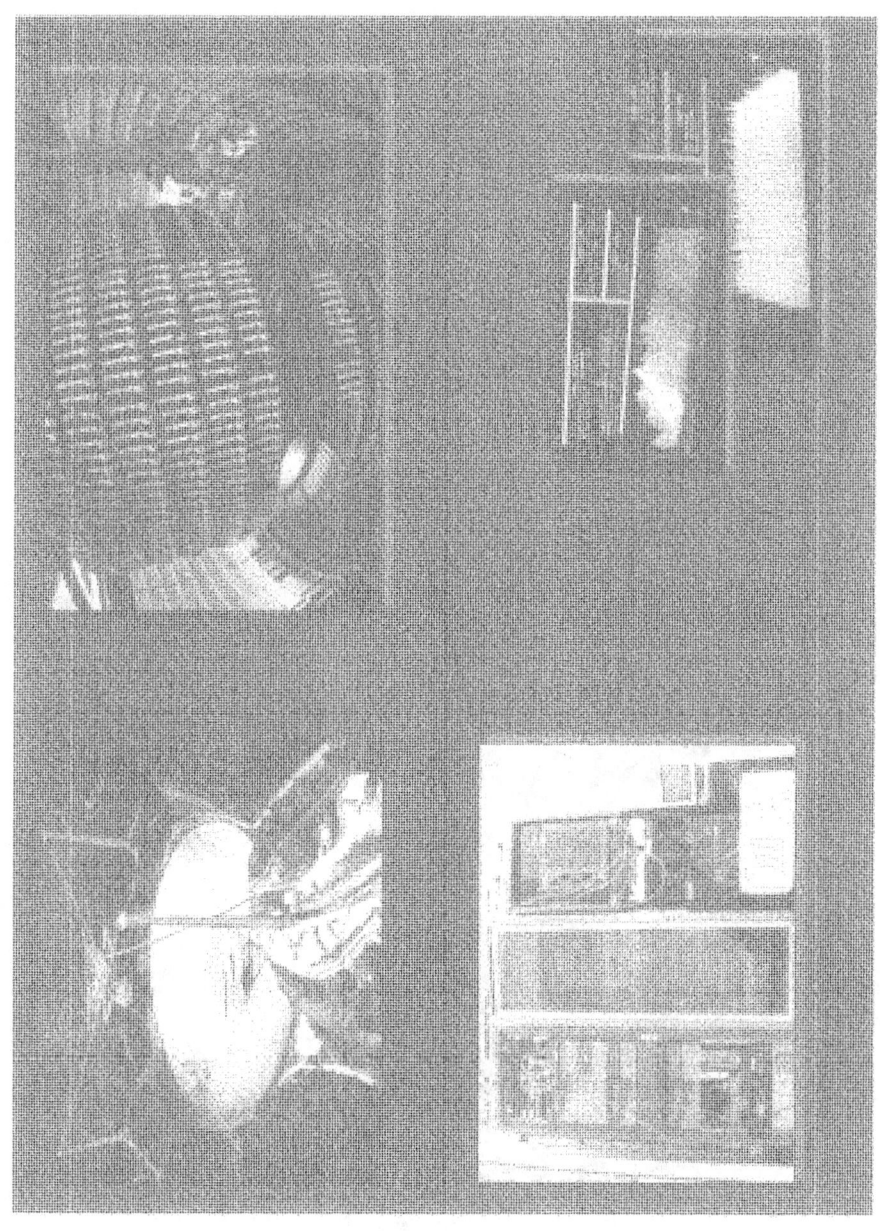

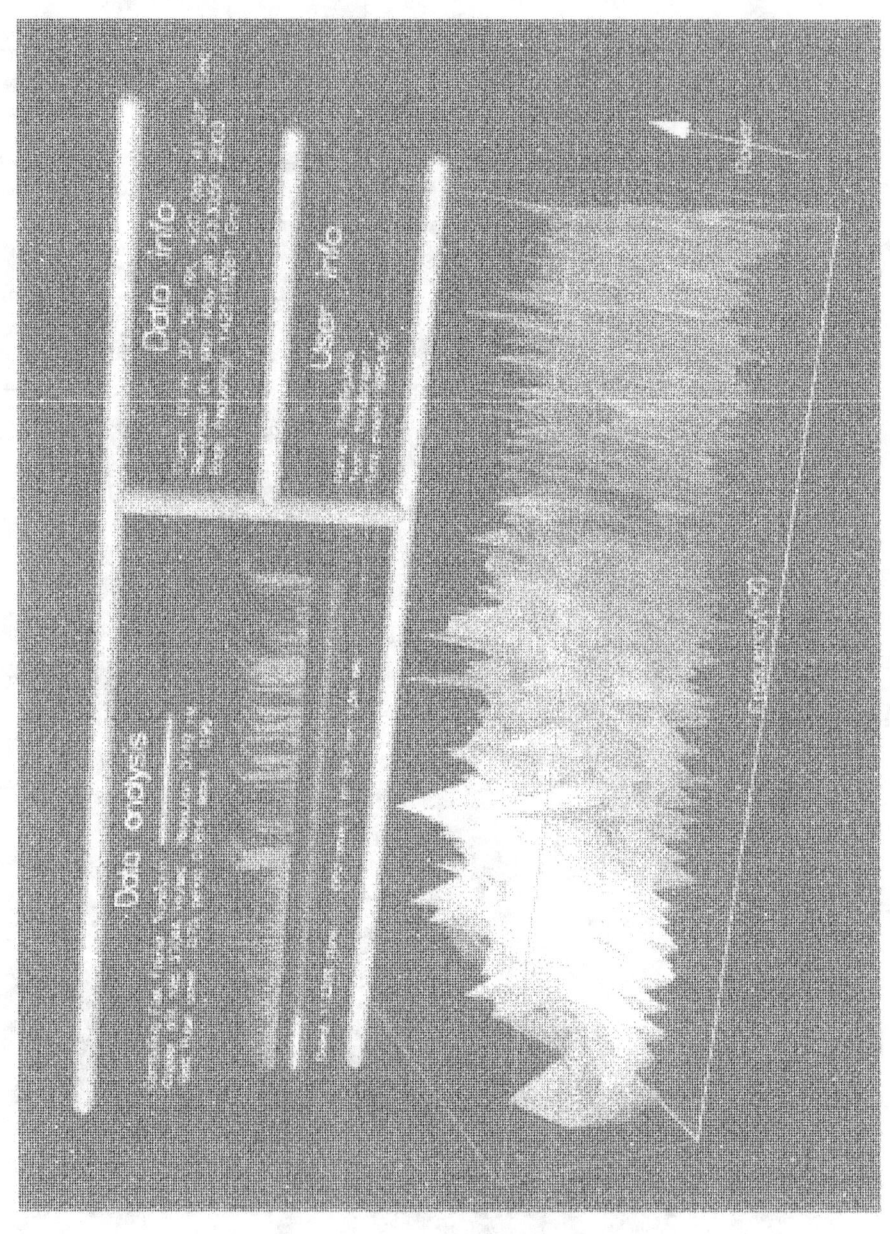

SETI@home Statistics

TOTAL   RATE

8,464,550   2,000 per day
participants
(in 226 countries)

3 million years   1,000 years per day
computer time

$3*10^{21}$ 1,000 Tera-flops
operations

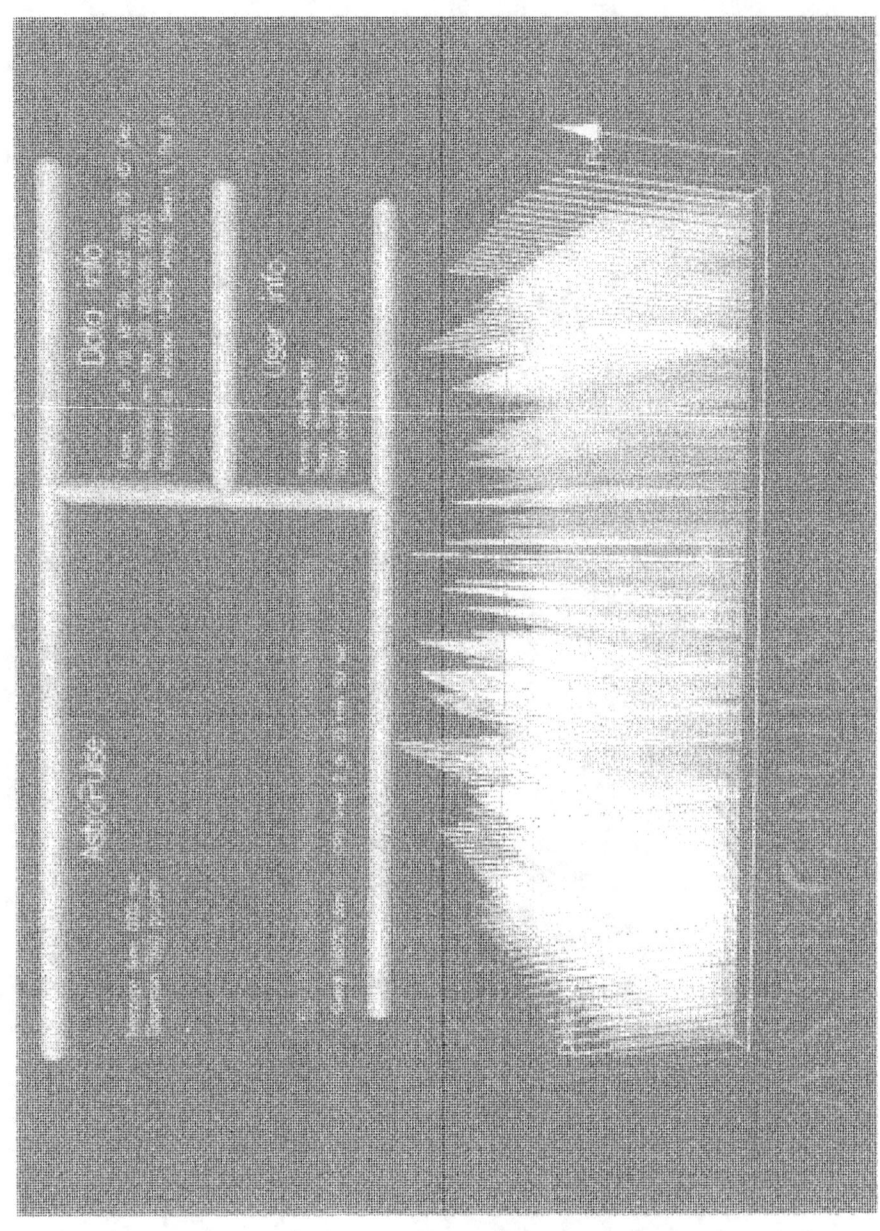

41

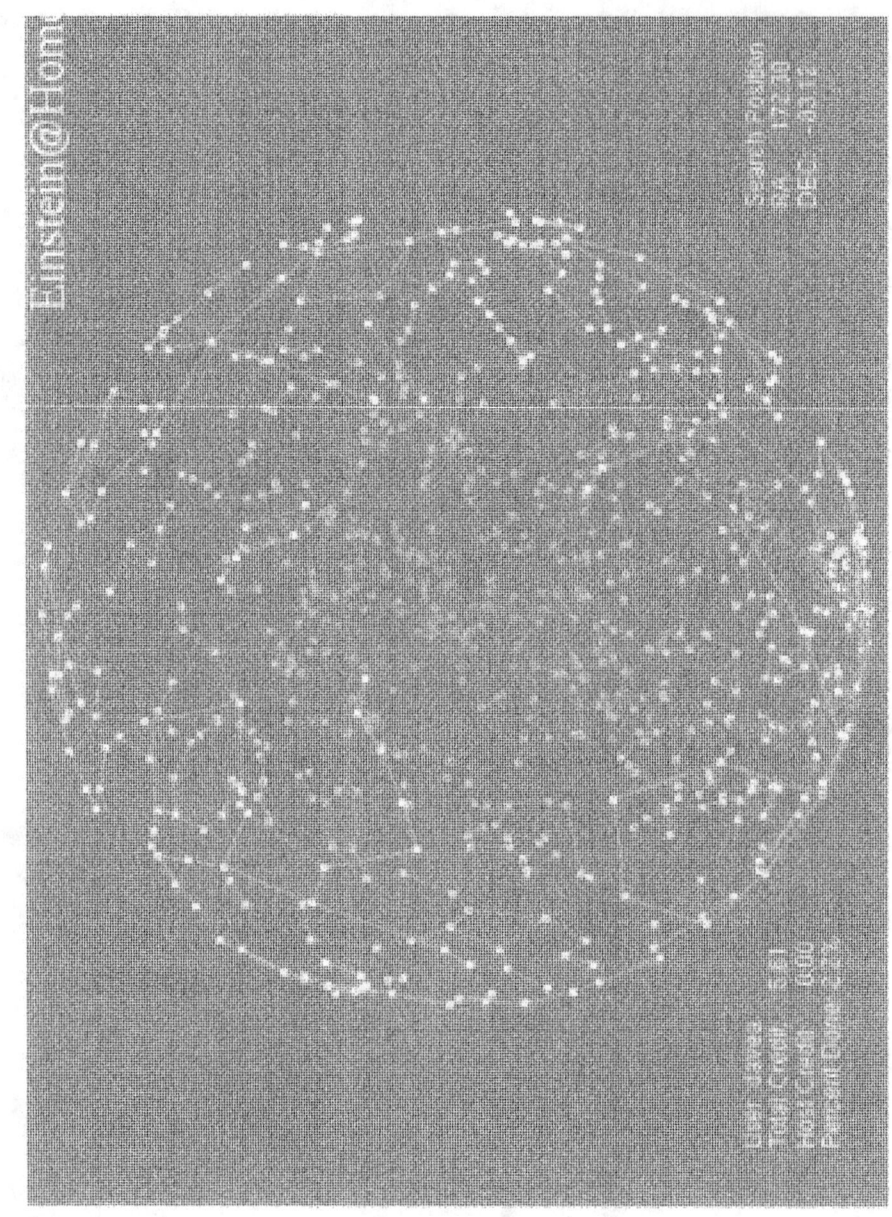

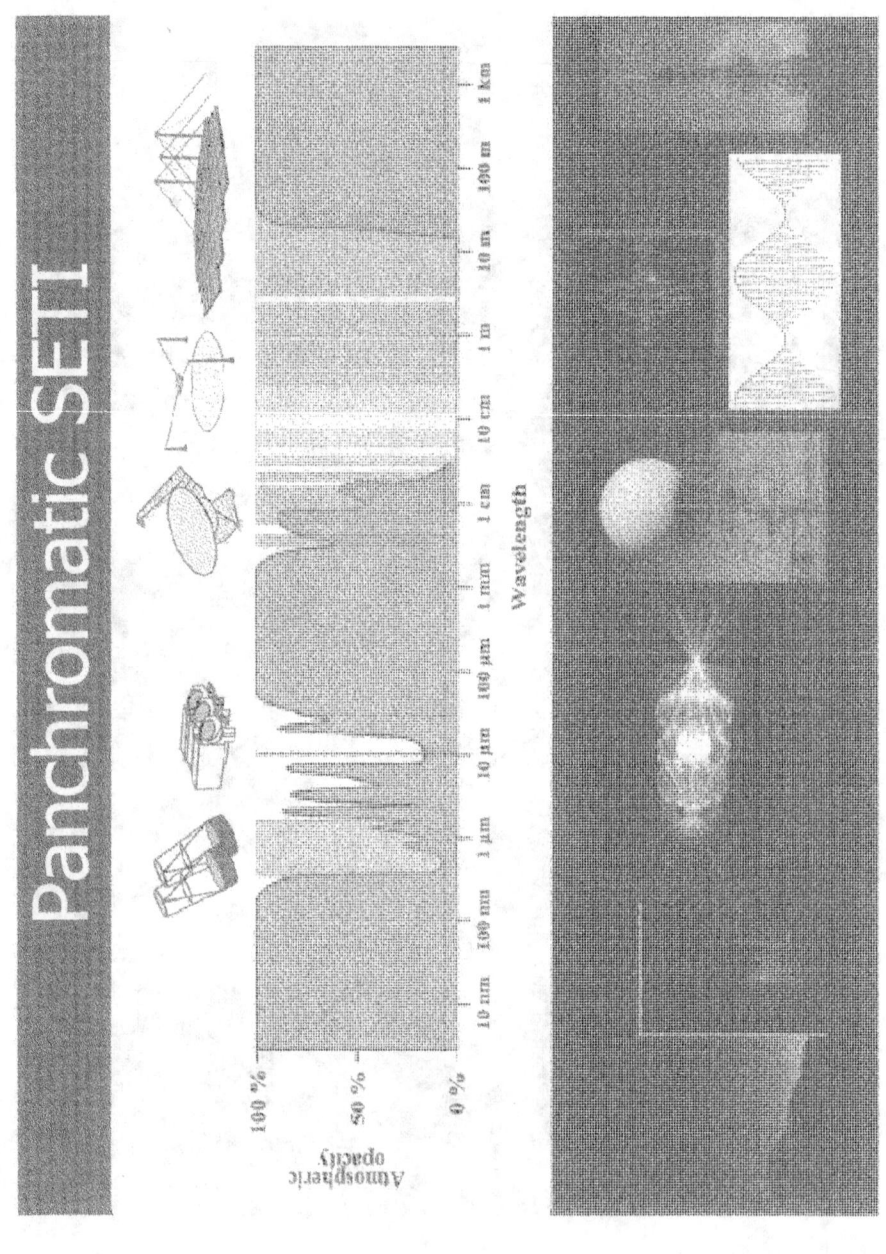

Panchromatic SETI

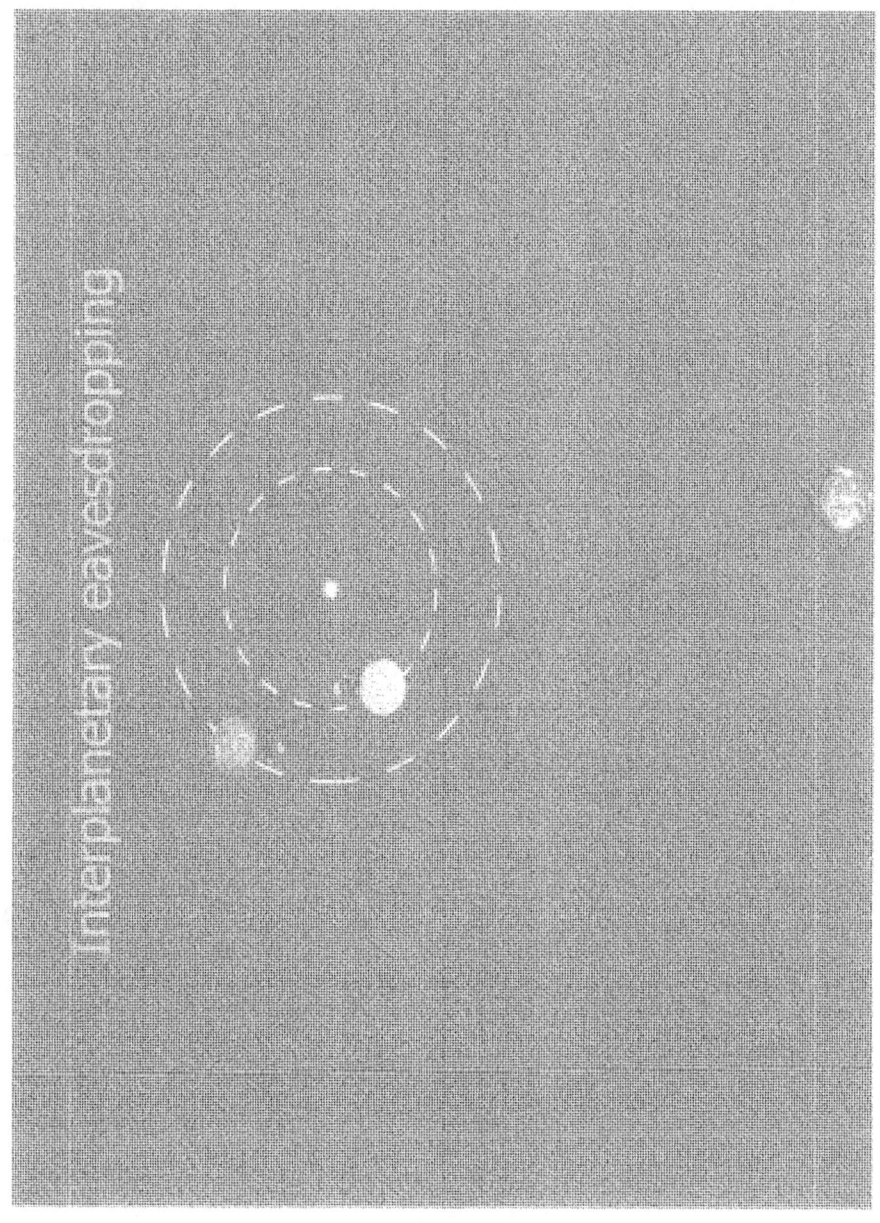

Interplanetary eavesdropping

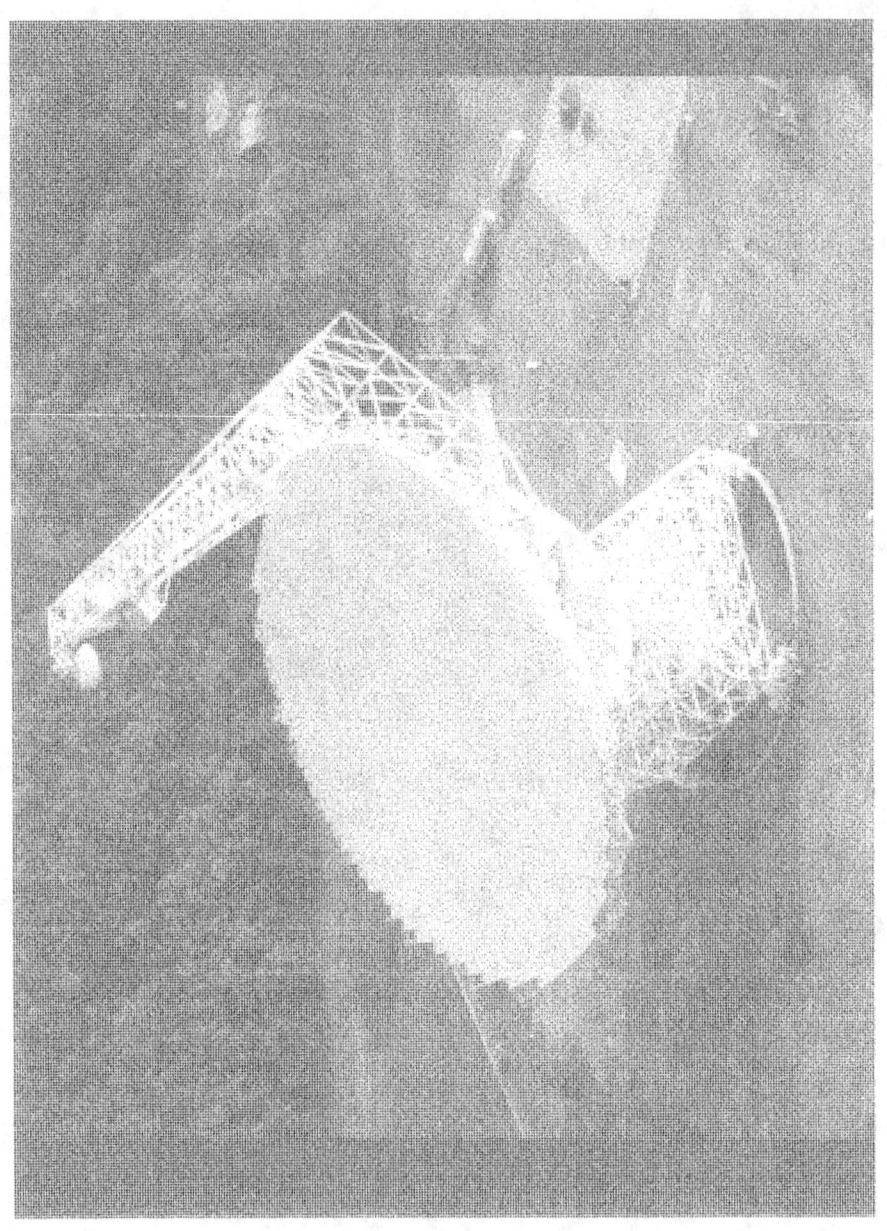

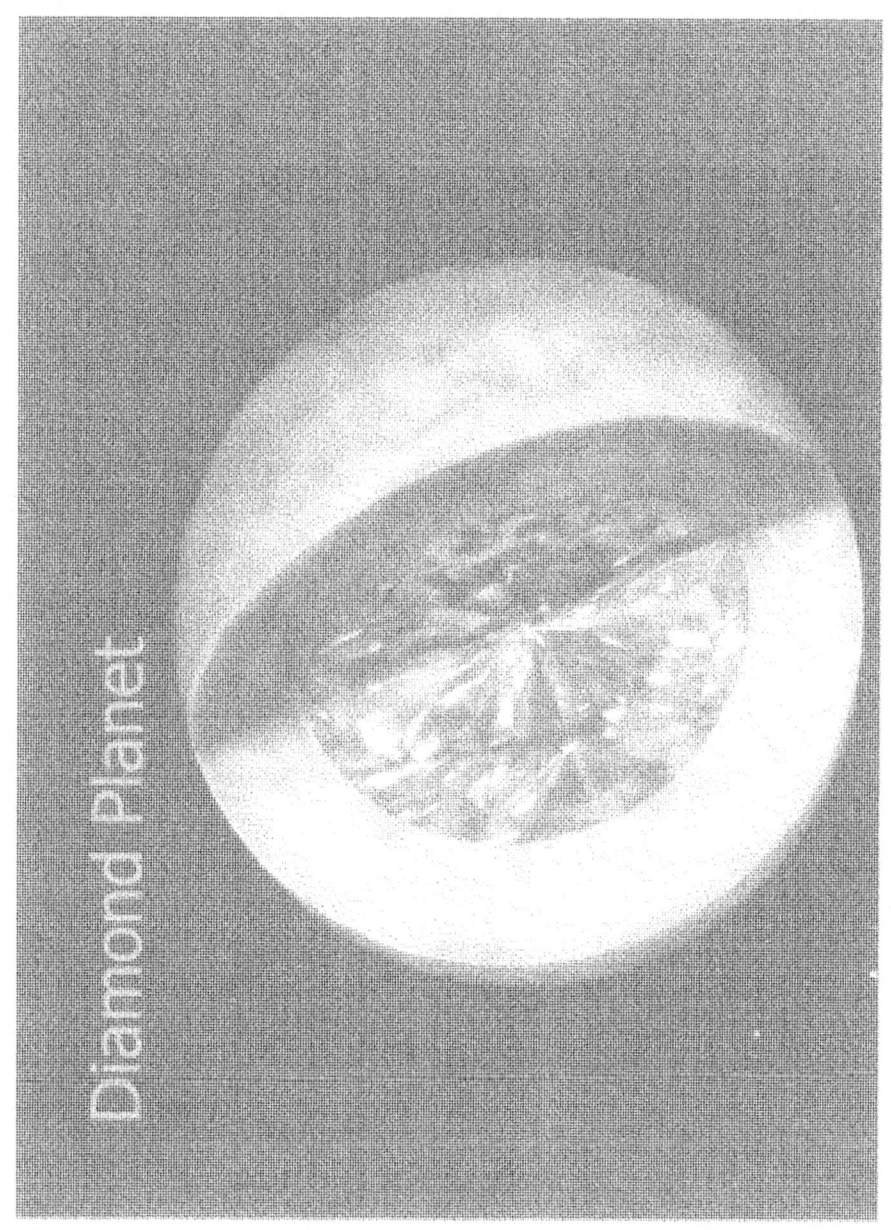

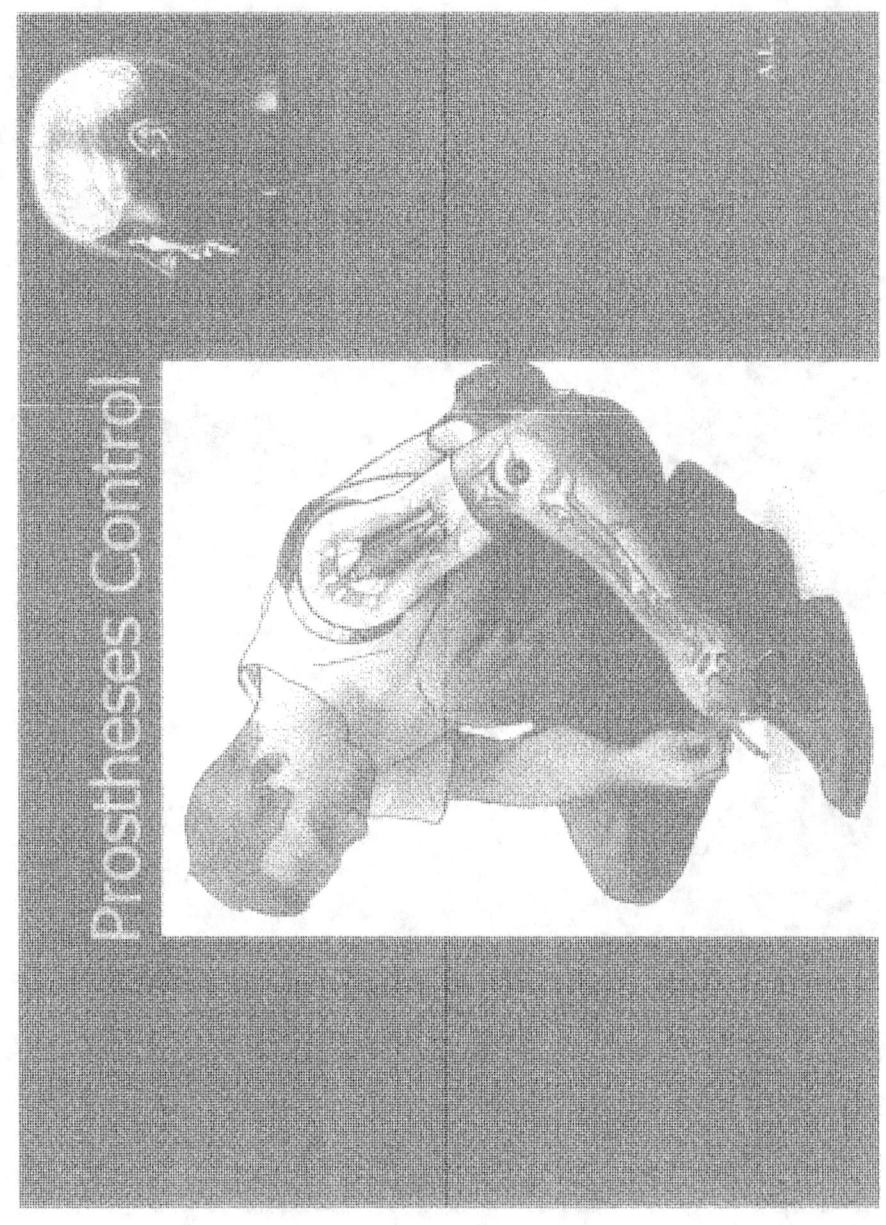

Prostheses Control

Summary and Conclusion

No ET so far

Still working on it

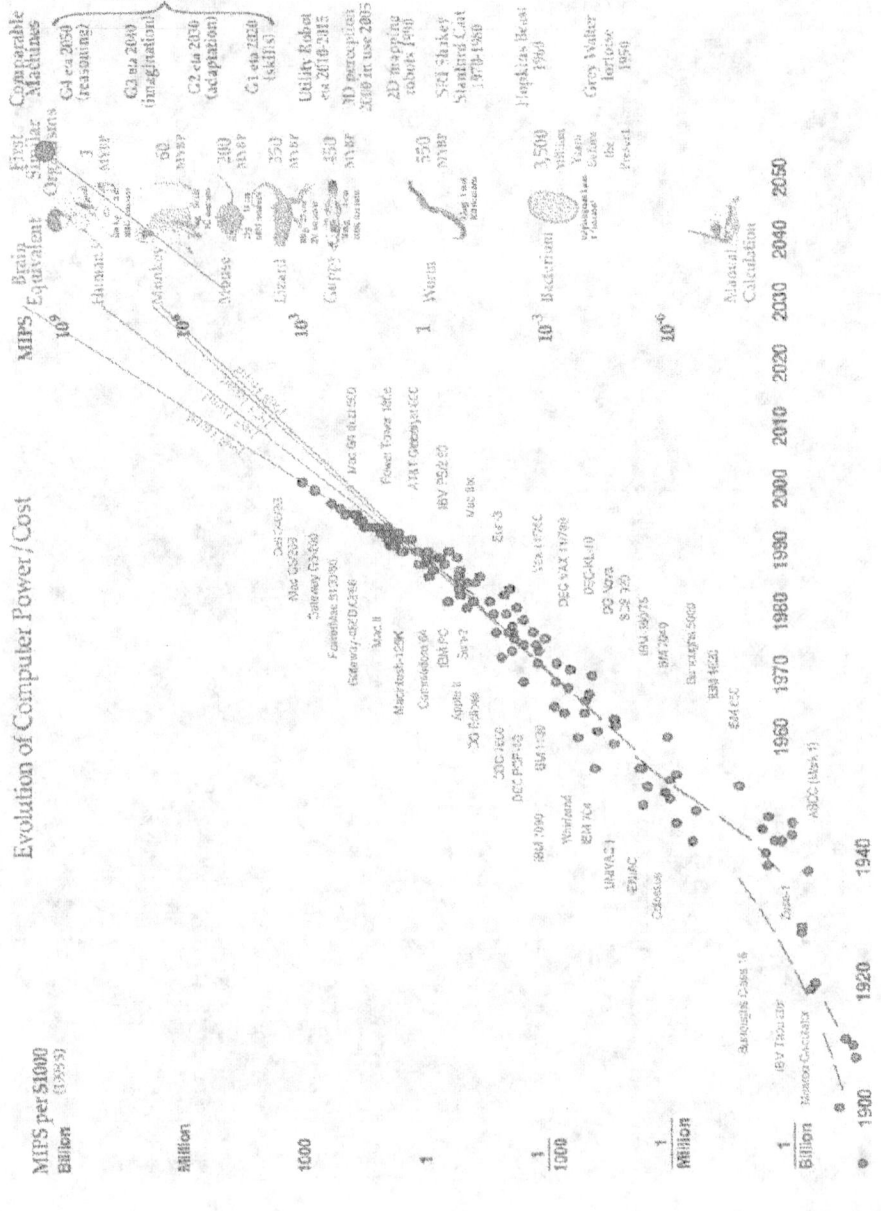

Evolution of Computer Power/Cost

SETI HAIKU

Searching for life
Answers are revealed
About ourselves

Paula Cook, Duke University

One million earthlings
Bounded by optimism
Leave their PC's on

Dan Seidner

Dan Werthimer   Biography

Short Bio

Dan Werthimer is SETI@home Chief Scientist and director of the SETI Research Center at the University of California, Berkeley. Werthimer also directs the Center for Astronomy Signal Processing and is associate director of the Berkeley Wireless Research Center. Werthimer was in the Homebrew Computer Club with Steve Jobs and Steve Wozniak. Everyone in the Homebrew club became filthy rich except Dan.

Long Bio

Dan Werthimer is SETI@home Chief Scientist and director of the SETI Research Center at the University of California, Berkeley. Werthimer also directs the Center for Astronomy Signal Processing and Electronics Research, and is associate director of the Berkeley Wireless Research Center. Werthimer was associate professor in the engineering and physics departments of San Francisco State University and a visiting professor at Beijing Normal University, the University of St. Charles in Marseille, and Eotvos University in Budapest. He has taught at universities in Peru, Egypt, Ghana, Ethiopia, Zimbabwe, Uganda and Kenya. Werthimer is co-author of "SETI 2020", editor of "BioAstronomy: Molecules, Microbes and Extraterrestrial Life" and "Astronomical and Biochemical Origins and the Search for Life in the Universe". Werthimer was in the Homebrew Computer Club with Steve Jobs and Steve Wozniak. Everyone in the Homebrew club became filthy rich except Dan.

Chairman SMITH. Okay. Thank you, Mr. Werthimer.

Thank you both for your excellent testimony, and actually you have anticipated my questions a little bit but I would still like to go forward with them.

And let me address the first question to both of you all but starting with Dr. Shostak. And it is this, kind of a two-part question. What do you think—and I can anticipate your answer a little bit on the basis of your statement—but what do you think is the possibility of microbial life being found in the universe or intelligent life being found in the universe? So the first question goes to the possibility. The second question would be what you think is the likelihood of finding either microbial life or intelligent life in the universe, two different kind of questions, Dr. Shostak.

Dr. SHOSTAK. Well, the probability of life of course is hard to estimate because what we do know now and something we didn't know until very recently, even 10, 20 years ago, is that there are habitats that could support life. What astronomy has proven in the last 500 years is that the entire universe is made out of the same stuff, right. The most distant galaxies have the same 92 elements that were on the wall in your 9th grade classroom. So this means that if you have taken chemistry in school you don't have to take it again if you move to another galaxy. It is all the same everywhere.

We know that the building blocks are there. We now know that there are going to be plenty of planets where you have liquid water and an atmosphere, the kind of salubrious conditions that you have in Hyattsville, for example, so that life could arise on any of these places.

We also know that life began on Earth very, very quickly. Now this is only a sample of one, so it is not entirely convincing, but it does suggest that it wasn't very difficult for life to get a foothold on this planet, and maybe elsewhere. So life I think is perhaps not so hard to get started. That is sort of the general impression among scientists. But what they believe is not so important; it is finding it that is important.

The second part of your question, what about intelligent life, that is a lot harder. The Earth has had life we know for at least 3.5 billion, probably 4 billion years, almost since the beginning. This place has been carpeted with life and almost all of that time it required a microscope to see it. It was all microbial. Only in the last 500 million years did we get multicellular life, eventually trilobites, dinos, you know the whole story, okay.

That opens up the question, well, you know, if I give you a million worlds with life, what fraction of them is ever going to cook up something as clever as you all? And the answer to that is we don't know the answer to that. However, there are indirect suggestions that it will happen given enough time simply because we are not the only species that has gotten clever in the past 50 million years. If you have dogs and cats at home, they are cleverer than the dinosaurs. Intelligence does pay off.

Chairman SMITH. Thank you, Dr. Shostak.

By the way, you have made a point that I might emphasize and that is that 20 years ago we hadn't detected a single planet outside our solar system. Now, we are up to close to 2000 so it is almost exponential growth in astrobiology research.

Mr. Werthimer.

Mr. WERTHIMER. I suspect the universe is teaming with micro-bial life. It would be bizarre if we are alone but I don't know that for sure. The intelligence is going to be rarer, but because there are trillion planets, I believe it is going to happen often. It has happened several times on this planet and it is likely to arise else-where.

Chairman SMITH. And as you would put it, at 100 percent then?

Mr. WERTHIMER. 99.

Chairman SMITH. Yeah, 99.9999 and strung on out. Okay. Good. The next question, Mr. Werthimer, let me follow up with you. And by the way, as far as the SETI@home screensaver goes, that would be something for the students here to take advantage of as well as Members. I tried to adapt that to my laptop in my office several years ago and was not able to, so maybe we will talk some more. Maybe the government needs to change its policy; I am not sure which.

But let me ask you what are the advantages and disadvantages of radio SETI versus optical SETI?

Mr. WERTHIMER. There are lots of pros and cons. Lasers are good for point-to-point communication and lots of bits per second, lots of data. I think the best strategy is a multiple strategy. We should be looking for all kinds of different signals and not put all our money in one basket.

It is hard to predict what other civilizations are doing. If you had asked me a 100 years ago what to look for I would have said smoke signals, so we tried to launch a new SETI project, a new idea every year.

Chairman SMITH. Okay. And, Dr. Shostak, anything to add to the advantages or disadvantages of radio versus optical SETI?

Dr. SHOSTAK. I should point out that they are both sort of different colors of the same thing, in fact literally different colors because they are both electromagnetic means of communication and we use both in our telecommunications here on Earth and I suspect the aliens will as well.

I have to say that just about every week I get an email from somebody who says you guys are looking for radio signals? That is so old school. The extraterrestrials, assuming they are out there, will use something much more sophisticated than that and I am not sure what that is. That depend on physics we don't know. And one shouldn't discount a technology simply because it has been around a while. We use the wheel every day. That is a pretty old technology. I suspect we will continue to use the wheel for a long time.

Chairman SMITH. Okay. And thank you both for your answers to my questions.

And the Ranking Member Ms. Johnson is recognized for her questions.

Ms. JOHNSON. Thank you very much. I am trying very hard to ask something that sounds sensible.

What is the status of the extraterrestrial intelligence research now?

Mr. WERTHIMER. So I think we are just getting in the game; we are learning how to do this and I think we would be lucky to find—

even though I am optimistic about life and intelligent life in the universe and it is likely there is a whole galactic internet out there, I think we would be lucky to find them now but I am optimistic in the long run.

Dr. SHOSTAK. Congressman Johnson, I might point out that contrary to popular impression, this experiment isn't the same from day-to-day. People figure we are sitting around with earphones listening in to cosmic static every day, a rather tedious job if that is what it were. But it is not. All the listening is done by computers.

But the really important point is that much of this experiment depends on digital technology, computers if you will. And as you know, there is something called Moore's law which says that whatever you can buy today for a dollar you can buy twice as much for a dollar two years from now. There is this very rapid growth in the capabilities there.

So in fact this search is speeding up and it is actually speeding up exponentially, a very heavily overused word exponentially, but in fact it applies.

Ms. JOHNSON. Tell me this. I know that the improvement of technologies are important and yet some of the old technologies or old techniques are also still in play. How do you predict your advancement based on what you have available to you for research tools?

Dr. SHOSTAK. I will just say something. I am sure Dan has much to add to this. But in terms of that we can do in the near future, the foreseeable future, what you really I think need to do if you want to have a decent chance of success—and mind you, this has to remain speculative; I mean this is all like asking Chris Columbus two weeks out of Cadiz, you know, hey, have you found any new continents lately? And his answer would be, well, there was only water around the ship today, and by the way also yesterday water around the ship and tomorrow it is going to be fairly aqueous in the vicinity of the ship, okay. So he can't predict when anything interesting is going to happen, nor can we.

But if you look at what are called euphemistically estimates— and they are guesses—as to what fraction of stars out there that house somebody that you might be able to pick up, it sounds like you have to look at a few million star systems to have a reasonable chance of success. We can't do that today. We have not that today. We have done less than one percent of that as of today, okay. But given the predictable advancements in technology, to look at a few million star systems is something that can be done within two dozen years given, you know, the funding to do it.

Mr. WERTHIMER. Seth captured it well.

Ms. JOHNSON. Now, when we find the other life on other planets, what do you speculate we would find and what is of value or potential value?

Mr. WERTHIMER. I think it is profound either way. This is not an expensive thing, of order $1 million a year we are founded by National Science Foundation, NASA, Templeton Foundation, some private donations.

The reason I think it is profound either way if we discover that we are alone, we had better take really good care of life on this planet. It is very precious.

And the other thing that is profound, too, if we are—find that we are part of a galactic community and get on the galactic internet and learn all their poetry, music, literature, science, we could learn a lot.

Dr. SHOSTAK. I would just add briefly nobody knows what we will learn. If we can decode this signal, this is sort of like being confronted with hieroglyphics. You know, you might be able to figure them out. In the case of the hieroglyphics, it wasn't so hard. It turns out the hieroglyphics were written by humans so that made it a lot easier. And there was also the Rosetta Stone and whatever.

So we might not ever figure it out, okay. If you could, you would be listening to data being sent by societies that are far in advance of us because we are hearing them, not the other way around. So they are more advanced and maybe they teach you some very important stuff. Who knows? I mean imagine that the Incas find a barrel that is washed up on the shore, you know, maybe from Europe and it is filled with books. If they could ever figure out the books, they would learn a lot of interesting stuff. I don't know that we will ever figure out the books, but even if we don't, the important point has been made, and that is we have calibrated our place not in the physical universe, we have sort of done that, but calibrated our place in the biological and even more—the intellectual universe. And I think that that is maybe good for our souls to know how we fit in.

Ms. JOHNSON. Thank you very much. My time has expired.

Chairman SMITH. Thank you, Ms. Johnson.

The gentleman from Ohio, Mr. Johnson, is recognized for his questions.

Mr. JOHNSON. Thank you, Mr. Chairman.

Gentlemen, for both of you, how has the recent discovery of over 1,700 planets by the Kepler space telescope—how has that impacted SETI research?

Mr. WERTHIMER. If you had asked astronomers 20 years ago are there planets going around other stars, we would have said, well, we think so but we don't know. And that has all changed now. And a lot of it is due to NASA's Kepler mission. And if you extrapolate on the planets, which are a few thousand planets that they have discovered, if you extrapolate on that, there are a trillion planets in the Milky Way galaxy. That is about three or four times more planets that there are stars, so that has got a lot of places for life.

Mr. JOHNSON. Okay.

Dr. SHOSTAK. I think that it has also affected the experiments in the sense that in the past we would point the telescopes in the direction of stars, certain kinds of stars, certain masses of stars, certain brightnesses of stars. Those stars were the ones that we thought might have an Earthlike planet, but we didn't know. We now know two things. One, as Dan has just mentioned, we know that the majority of stars have planets. So you can just look at a random star and feel fairly confident that it has planets. But more than that, we are beginning to get some indication from Kepler what fraction of stars have planets that are sort of like the Earth, and that fraction is not one in a million, it is not 1 in 1,000, it is not 1 in 100. It may be one in five. So you look at, you know, 50 star systems and you have examined 10 Earthlike planets. So in some sense it has made the search much more straightforward. We just look at all the nearby stars we can.

Mr. JOHNSON. Okay. Well, Dr. Shostak, would you please provide some examples of the technical contributions that SETI has made to astronomy and other fields? For example, how has SETI research benefited other areas of science?

Dr. SHOSTAK. Well, I think that its benefit is less so in terms of the discovery. Obviously we haven't found ET. If we had, we wouldn't be having this hearing, okay. But—and to my surprise, I have to say SETI has not turned up any astrophysical phenomena that were unexpected as well, okay. And that is surprising. Normally, the history—the precedent in astronomy is that every time you build an instrument that examines a different if you will parameter in the phase space of the universe, you find something new. So it is instructive that it has not.

The kind of technology that has been developed is certainly of interest to other fields in astronomy. But I think the real value of SETI is not so much in terms of what it does to astronomy but what it does in terms of the other efforts being made to find life in space. NASA has a big effort. You know, the rovers on Mars, yes, they are there to find the hydrology, the history of water on Mars, but why are you interested in the history of water on Mars? You are interested because you want to know were there ever Martians, you know, microbial most likely, or are there still Martians? That is what interests people the most.

And SETI was always, if you will, a punch line to this story that NASA had about finding, you know, the traces of water on Mars or burrowing through the ice on Europa and Enceladus, some of these moons of the outer solar system where there may be vast quantities of liquid water, that sort of thing.

SETI was always that, okay, life— we may find life, but what about intelligent life? That would be even more interesting. And that is what is missing from the NASA program today.

Mr. JOHNSON. Okay. You made a comment just a few minutes ago that kind of caught my attention. Let me make sure I got it right. You said that if we hear from intelligent life out there somewhere that they must be more advanced than us because we are hearing from them and not the other way around. How can you draw that conclusion? I mean maybe they have been hearing from us for a long time and just don't like what we have to say.

Mr. WERTHIMER. Um-hum. I think it is entirely possible that we are on their—in their catalog. They have seen oxygen in our atmosphere and they know we are out here. And I think life in the universe is going to be—there is going to be lots of different stages. Some of it is going to be microbial, some of it will be trees, more sophisticated. The Earth is 5 billion years old, some stars are 10 billion years old, so there could be a lot of advanced civilizations as well.

Mr. JOHNSON. Um-hum. Okay.

Dr. SHOSTAK. Just to point out that you are not going to hear from any less advanced societies. They are not building radio transmitters.

Mr. JOHNSON. Well, yeah, I would say——

Dr. SHOSTAK. That is for sure.

Mr. JOHNSON. I would say at least equal to, perhaps more advanced, but, you know, maybe they got their caller ID block turned on or something.

Dr. SHOSTAK. It could be. I wouldn't speculate on alien sociology and whether they would like our television or not so I don't know about that. But the chances that if they are at least at our level that they are within 100 or even 1000 or even 10,000 years of our level is simply on statistical grounds highly uncertain. If you hear from somebody, they are way beyond you.

Mr. JOHNSON. Yeah. One final quick question for both of you. How would you define successful SETI research? I mean I know that is kind of a nebulous question but——

Dr. SHOSTAK. Finding the signal.

Mr. JOHNSON. —how would you define it successful?

Dr. SHOSTAK. If you found a signal that could be corroborated. If you just find it once and you can't find it again, it is not science. So if you find a signal that is moving across the sky the way the stars do because of the rotation of the Earth, it is a narrow band signal, it is not made by nature, it is made by a transmitter, that is success.

Mr. JOHNSON. Right. Okay.

Mr. WERTHIMER. I think the most likely scenario is finding some sort of artifact of their technology, a radar signal or a navigational beacon or something. That won't contain a lot of information but we will know we are not alone.

Mr. JOHNSON. Okay. Thank you, Mr. Chairman. I yield back.

Chairman SMITH. Thank you, Mr. Johnson.

The gentlewoman from Oregon, Ms. Bonamici, is recognized for her questions. And if the gentlewoman will just yield to me for 10 seconds.

Ms. BONAMICI. Certainly.

Chairman SMITH. It was mentioned a while ago that the likelihood is if there were other intelligent civilizations, they would likely be far and—more advanced than we are. We are a relatively junior galaxy. They might be two—I don't know, two billion years older than we are and it is just fascinating to think what form of life might be existent in a universe or a parallel universe or another galaxy where they have had a two billion year head start. We might not even recognize the sentient beings. We might not be able to communicate with them, but that is just one of the reasons why we are fascinated by the subject.

And none of this will be counted or charged against the gentlewoman's five minutes for questions.

Ms. BONAMICI. Thank you, Mr. Chairman.

Thank you so much, Dr. Shostak and Mr. Werthimer, for being here. I noticed in your testimony, Mr. Werthimer, that you said that there are 24 SETI scientists on the planet and I can't think of a time in this Committee were we have had a larger percentage of experts on our panel. So thank you both so much for being here. I really appreciate it.

And, Dr. Shostak, I really am intrigued by your section in your testimony on the public's interest and how the idea of life in space is an idea that everyone grasps and is especially an ideal hook for interesting young people in science. I think that is evidenced by the full Committee room today.

One of the statements that resonated with me is "it would be a cramped mind indeed that didn't wonder who might be out there." I really appreciate that. You said also in your testimony "extraterrestrials are the unknown tribe over the hill, potential competitors or mates, but in any case, someone we would like to know more about." And I recollect a similar hearing in this Committee. I believe it was last year when one of my colleagues—and I am fairly certain it was Representative Chris Smith, who is no longer on the Committee, said the interesting question is what do we do when we find the life on another planet?

So can you talk, both of you, about what is the plan? Do we announce it to the world? Do we do research more to determine if these are friendly or collaborative? Or what do we do when we make the discovery, assuming that it is going to happen?

Dr. Shostak, would you like to begin?

Dr. SHOSTAK. Yes. That is a question of great interest to the public and of great importance as well. To begin with, there is no danger. You tune in your favorite DJ here in D.C. on the car radio and there is no danger that that DJ is going to jump into the car next to you and give you a hard time because they don't know that you have picked them up. So if we pick up a signal, they don't know that.

There is the question of, well, should we reply? I will get to that in just a second. But what happens then? Suppose we do pick up this signal? It would be announced. The public has the idea that you all have a secret plan, that the government has a secret plan for what to do if we pick up a signal. As far as I can tell, there is no plan, okay. And we have had false alarms and I have waited for my Congressman to call me up and say, hey, you guys are picking up a signal. What about that? And nobody in the government shows the slightest bit of interest to be quite honest. What happens is that the media start calling up, the New York Times will call up, right, but the media—or, sorry, the government is not so interested.

So what would happen is that it would immediately be known that we had found this signal and it would be known even before it had been corroborated. So there are going to be false alarms. Be prepared for that. But what you do is you get somebody in another observatory to also observe it. You would not believe it yourself if you were the only ones to find it. There are too many things that could go wrong, okay.

Ms. BONAMICI. Mr. Werthimer, do you have anything to add to that?

Mr. WERTHIMER. Yeah. I think before you make a big announcement you want to make sure it is real. You ask a different telescope with different people, different software, different equipment to see if they can verify it. Then you can triangulate, make sure it is coming from something outside. You make sure it is not a graduate student playing a prank on you. And once you have some confidence that you have found something, you may not know what it is. It could be some new astrophysical phenomena. When pulsars were discovered, they thought maybe they had found little green men. So I think you—then you—at the point where you are pretty confident that you have found something, you make all the information public, the coordinates in the sky, the frequency, anything you know about the signal, and then I think there will be a lot of debate about whether there is some new natural phenomena or this is really evidence of another civilization. A lot of people will be working on that problem.

Ms. BONAMICI. And could you also address of the 24—you say the 24 SETI scientists on the planet, to what extent are other nations involved? How collaborative are we? We have a lot of discussions in this Committee about international collaboration, especially in space. So can you talk about where we are as a nation compared with the other countries in the world——

Mr. WERTHIMER. Yeah.

Ms. BONAMICI. —on this issue?

Mr. WERTHIMER. SETI is quite fragile. As you said, there are 24 people doing it. There are about two thirds of them in the United States. The United States is leading this effort and a lot of the original ideas have come out of the United States. But there is— we are working with other scientists in other countries, and because it is so fragile, we are trying to train new people and get new ideas and get other groups because it is only at a very small number of institutions right now. The funding is fragile, too. It is fluctuating around.

The two biggest telescopes on the planet are currently funded by the National Science Foundation, the Green Bank telescope in West Virginia, the Arecibo telescope. Those are in funding jeopardy. It looks like one of those observatories is probably going to have to be shut down. The other is just hanging by a thread. The Chinese are building a bigger telescope. There is a new one going to be built in South Africa and Australia. So the United States may not continue to lead this work but it is now.

Ms. BONAMICI. I would find that disappointing if that happened.

And then I am out of time. I yield back the balance of my time. Thank you, Mr. Chairman.

Chairman SMITH. Thank you, Ms. Bonamici.

And the gentleman from New York, Mr. Collins, is recognized for his questions.

Mr. COLLINS. Thank you, Mr. Chairman.

I think I might ask the question everyone in this room wants to ask. Have you watched Ancient Aliens and what is your comment about that series?

We will start with you, Dr. Shostak.

Dr. SHOSTAK. Yes, I think I have been on it actually, more than once. The public is fascinated with the idea that we may be being visited now or may have been visited in the past, the so-called UFO phenomenon. I personally don't share the conviction that we are being visited. I don't think that that would be something that, you know, all the governments of the world had managed to obfuscate, to keep secret. I don't think—I don't believe that.

But the idea that maybe we were visited during the time of the ancient Egyptians and so forth, keep in mind that in the 4.5 billion year history of the Earth the time of ancient Egyptians was yesterday, right. So, again, why were they there then? What was it that brought them to Earth? I have no idea and I don't find very good evidence. I don't think—I think the pyramids, for example, were probably built by Egyptians. I know that that is a radical idea for some people but the Egyptians were very clever and they could certainly do that.

So I don't think that there is any good evidence that convinces me that we were visited in historic times.

Mr. COLLINS. How about you, Mr. Werthimer?

Mr. WERTHIMER. UFOs have nothing to do with extraterrestrials, so even though I am optimistic with life, there is no evidence that any of these sightings—I think some of these sightings are real phenomena. We get a lot of calls when the Space Station goes over, although some people embellish and they say it has windows and things. And some of it is people's imagination and we know that because it ties very closely to popular culture. When Jules Verne wrote about flying saucers, everybody sees—started seeing flying saucers. Before that, people saw angels. When people watch movies, then we get a lot of reports that are tied to what is in the movies. And some of it is actually deliberate hoaxes, you know, for people making money.

Mr. COLLINS. Yeah. Thank you. I think that was my only question, Mr. Chairman. I yield back.

Chairman SMITH. Thank you, Mr. Collins.

The gentlewoman from Maryland, Ms. Edwards, is recognized for her questions.

Ms. EDWARDS. Thank you, Mr. Chairman. I feel like I should have been here earlier so I apologize. I have enjoyed the discussion thus far and reading the testimony.

You know, my favorite movie is Contact, right, so every year it comes out since 1997 I watch it. I dream. I think, well, you know, who knows? What is intriguing about this conversation is the idea that—and it is a little bit of hubris, right, that somehow we are waiting to find them as opposed to them finding us. And maybe that is just the nature of Homo sapiens. That is kind of what we do.

But I am a little bit curious. Dr. Werthimer, in your prepared statement you discuss the panchromatic SETI project, which will use six telescopes to search nearby stars and stars most likely to host an exoplanet system similar to the sun's. And so the project as you described it would examine a large portion of the electromagnetic spectrum spanning from low frequencies through optical light to detect possible signals from advanced civilization. How are the target stars that you have talked about identified and how are you going to coordinate the use of the six telescopes?

Mr. WERTHIMER. We are not trying to use the telescopes all at the same time. That is actually hard to do so we just—we use a telescope. And other groups are—we are working with a lot of groups at universities and observatories. But typically, we will use one telescope and then a month later we will use another telescope, and so on.

The stars that we are targeting, we—instead of targeting stars that we know have planets because of Kepler spacecraft, it looks like all stars have planets, so we are just going to target the nearest stars. So that is our plan is just target the nearby stars.

Ms. EDWARDS. Great. And you talked also about this notion that there are just sort of 24 of you folks most interested robustly, academically studying this, but aren't there like a whole—there is a whole network of people out in communities who kind of feed or fuel some of the research that you are doing?

Mr. WERTHIMER. Seth, you want to take that one?

Dr. SHOSTAK. Dan refers to me because I don't think we know the answer to that question. In order to do this, it would be like saying, you know, sure, there are a few thousand people looking for the Higgs Boson but what about the communities that are feeding that? If you don't have the instrument, it is very hard to do the experiment. And the number of instruments involved here is very small.

Ms. EDWARDS. So the rest of us are really just, you know, dreaming and pretending that that is what we are——

Dr. SHOSTAK. Well—

Ms. EDWARDS. That is all right. You don't have to answer that. I was not serious at all. And then I want to talk about security issues in the time that we have left.

I understand that early on there was an assessment of the robustness of the SETI's home software to withstand malicious attacks and penetrations. And in the earlier study you found that there had been two noteworthy attacks and the web server was compromised. And you also found later that exploiting a design flaw in your client/server protocol that hackers had actually stolen thousands of user email addresses. Can you give us an idea of the current state of security?

Mr. WERTHIMER. Yeah. I think in general downloading software and installing it on your computer, you should be careful. It actually turns out that SETI@home is one of the safest things you can install on the computer, and the reason is because millions of people are using it and testing it out, and so—and also it has been running for a really long time and it is open source software. The software is—anybody can read the software and help us—a lot of the volunteers actually help us write the software and we are now reporting it to cell phones so you can run it on a cell phone, which will allow us—a lot of people—even more people to participate in the search.

Ms. EDWARDS. I guess some of the question is just the—when—especially whenever you deal with open source, the challenge of the system's vulnerability.

Mr. WERTHIMER. Yeah. I actually think open source software is actually a little safer because so many eyeballs can look at it.

Ms. EDWARDS. Okay. I am done. I think I will just go back to watching my movies.

Chairman SMITH. Thank you, Ms. Edwards.

The gentleman from Florida, Mr. Posey, is recognized for his questions.

Mr. POSEY. Thank you, Mr. Chairman. And thank you for inviting these distinguished witnesses for this fascinating testimony, very enjoyable.

I go to the SETI Facebook page every day to get a little extra factoid, learn something every day. I hadn't been there a single day to find that I already knew your message of the day, very educational, very inspiring, obviously very interesting, and the graphics are always good, too, and I want to thank you for that.

On your disclosure I was really impressed with the number of agreements and grants. I am just really glad to know that NASA is so engaged with what you are doing there and still allow you all to have a pretty free hand to do what you do, better I think than anybody else is doing it obviously. And so thank you for that.

Obviously there is some curiosity about your thoughts about such things as Project Blue Book. What do you think?

Dr. SHOSTAK. First off, I want to thank you for noting all those grants, by the way, are for the astrobiology research being conducted at SETI Institute. There is actually no federal money going to the search for intelligent life.

Mr. POSEY. Right.

Dr. SHOSTAK. But we do—the majority of our scientists are doing astrobiology, so life on Mars, the outer solar system. In terms of—

Mr. POSEY. And we are glad you are.

Dr. SHOSTAK. Yes. Well, so are we. I can assure you. And that is, I think, a very productive line of research as well.

In terms of Project Blue Book and the whole UFO phenomenon, I am personally quite skeptical. One-third of Americans believe, as I say, that we are being visited. That is the result of polls that have been taken since the 1960s. That number doesn't change. And by the way, if you think this is an especially American opinion, that is wrong. One-third of Europeans, Australians, Japanese, and so forth believe that we are being visited. I do not. I honestly do not. I don't think that the evidence is very good. I think that if we were being visited, it would not be controversial. It has been, what, 60-some years since Roswell, for example. If you had asked the residents of Massachusetts 60 years after Columbus do you think you are being visited by Spaniards, that would not be controversial.

Mr. POSEY. Yeah.

Dr. SHOSTAK. I think that if they were really here, everyone would know that.

Mr. POSEY. Okay. Very good. Stephen Hawking, I believe, made some comments about contact with extraterrestrials or other life. Your thoughts about his comments?

Mr. WERTHIMER. Yeah. So this is a controversial topic about whether we should transmit messages. That is called active SETI, or METI, messages to extraterrestrial intelligence. Most people in the field think that we are just an emerging civilization and the first experiments we should do is just listening, trying to receive signals and see what is out there. We think that advanced civilizations are going to be peaceful if you watch Star Trek, but we don't know that and that may be naive. So my feeling is that we should be just listening for now and maybe in 1,000 or 10,000 years if we don't hear anything, we should think about transmitting signals. But that is a question for all humanity. It shouldn't be just up to a few scientists. And so that is a big decision about who should speak for Earth. So right now I think we should be listening and that is—I believe that is what Hawking would say as well.

Dr. SHOSTAK. I am going to disagree a little bit with my colleague here, Dan. I think that there is very little danger in transmitting, and if there is, we are already doing it. Yes, we are not deliberately targeting the stars in general, although we have done that in the past. NASA sent a Beatles song in 2008 I believe it was to the North Star. And it will take 450 years to get there and they may or may not like the Beatles but, you know, they used a fairly powerful transmitter. But the most powerful transmissions are coming off the airports, right, for navigation, for the DEW Line, all these things. These signals are on their way into space. They have already reached several thousand star systems. Any society that has the technical competence to threaten you across dozens, hundreds, thousands of light years of space, any society at that level can pick up these signals. So if you are really going to worry about this, you better shut down all the radars at the local airports, and personally, I don't think that would be a very good idea.

Mr. POSEY. Okay. And briefly, still related, your thoughts on thorium.

Mr. WERTHIMER. I am sorry. I am not familiar with the topic.

Mr. POSEY. Thorium—

Mr. WERTHIMER. Are you talking about nuclear—

Mr. POSEY. Yes.

Mr. WERTHIMER. —reactors—

Mr. POSEY. Yes.

Mr. WERTHIMER. —on thorium?

Mr. POSEY. Yeah.

Mr. WERTHIMER. I am really not an expert. I am sorry.

Dr. SHOSTAK. Only this, if you are talking about powering spacecraft—

Mr. POSEY. Yes.

Dr. SHOSTAK. If you send spacecraft to some of the more interesting parts of our solar system, they are in the boondocks of the solar system, out Jupiter, Saturn, and so forth. When you get to Saturn, the amount of sunlight has dropped by a factor of 100 so you can't use solar cells very effectively out there. You have to power the craft some way. I wouldn't worry too much about radioactivity in space of course because space has plenty of radioactivity. That is the nature of the cosmos, right. But if you are worried about the fact that these launches could go awry and that you would land these things on Earth, yes, that is a danger, but of course people are aware of that danger and they try and to mitigate.

Mr. POSEY. Thank you, Mr. Chairman, and thank both the witnesses.

Chairman SMITH. Thank you, Mr. Posey.

The gentleman from Arizona, Mr. Schweikert, is recognized for his questions.

Mr. SCHWEIKERT. Thank you, Mr. Chairman, and to our witnesses.

So, let's see, what have we learned so far? We have learned there is a chance that aliens don't like the Beatles, which I have trouble imagining, and they don't like our television programming, and there was a couple other things, oh, yeah, and Contact is the best movie, right? Somehow I thought that would be funnier.

A couple mechanical questions I just want to sort of get my head around some of the current scientific understanding. Let's walk through a scenario and you tell me if it is plausible or this is current thought. Asteroid hits the world, you know, hits our Earth, and rock is thrown out into, you know, the stellar, it carries DNA. Does that DNA survive? Doctor?

Dr. SHOSTAK. Yes. This idea is known as panspermia, and I am sure you are aware of that, the idea that one world could infect another world has been looked at. People have actually simulated the environment of space and put some of our earthly bacteria into a rock and put it, as it were, in space to see how long they could survive, for example. You know, would the DNA still be viable when it got someplace interesting? And the results, as I understand them, suggest that yes, if you are talking about, you know, communicable disease if you will within the solar system, could a rock from Mars have ceded the Earth, that is possible. There is no evidence that that occurred but that is possible. The life would survive. It would remain viable over the kind of timescales to send rocks in the solar system from one world to another.

But if you are talking about seeding worlds in other solar systems at the distances of the stars, the problem is space is a pretty harsh environment even for a rock because there is a lot of radiation and it is incredibly dry, so anything that is in there is going to be suffering desiccation for maybe hundreds of thousands, millions really—

Mr. SCHWEIKERT. Yeah.

Dr. SHOSTAK. —of years before it gets there. And the general consensus that I have heard is that it won't be viable when it does.

Mr. SCHWEIKERT. Count on that because I think that is the current sort of thought right now.

Mr. WERTHIMER. Yeah, so as you know, asteroids have hit the Earth many times and so it will be a really interesting question if life is found in our own solar system, like, for instance, Europa, which is a moon going around Jupiter, has a liquid ocean, there could be something swimming around down there. By the way, when I—I talk to elementary schools and I ask them how are we going to get through the ice and see if there is something swimming around down there? The boys all say we should use machine guns and bombs and the girls say we should melt our way through using mirrors, a little different. But anyway—

Mr. SCHWEIKERT. Once again proving there is something in our DNA which is different.

Mr. WERTHIMER. So, if we do find life in our own solar system, it would be really exciting to figure out is it exactly the same kind of life? Does it use the same DNA, the same amino acids, the same nucleotides? Is it identical chemistry? That would mean that rocks are going back and forth between these moons and planets in our own solar system, and it really happened in one place and was carried back and forth, as Seth was talking about.

That is not very interesting. What would be much more interesting would be discovering life that is different with a different chemistry because if we do find something like that on Europa or another moon or Mars, that means that the universe is teeming with life. If we can find it in two different kinds of life in our own solar system, that means there is a lot of life out there.

Mr. SCHWEIKERT. Yeah. It makes the imagination wonder.

Earlier, the Chairman—and I mean this with all the love in the world—was trying to say give me a percentage of life out there in existence. I remember doing this sort of as a sort of thought process with one of my professors many years ago. And I guess one of the mechanisms was from the beginning until today Earth has had 100 billion species or something of that and how many can do higher math and sort of give you sort of a—and we would use that as sort of a benchmark to try to do those calculations. And I guess our understanding was it is unknowable, you know——

Mr. WERTHIMER. Yeah.

Mr. SCHWEIKERT. —of what is out there, what isn't out there. I mean, you know, we see the world of large numbers, large planets, you know, these huge numbers.

Mr. WERTHIMER. Um-hum. On Earth intelligence has arisen several times independently. There are a lot of intelligent creatures, although none is quite as intelligent as us maybe. We are not sure.

Mr. SCHWEIKERT. Well, we always used the higher math as the——

Mr. WERTHIMER. Yeah. But I—my guess is that on some planets there are going to be selective pressures that select for different kinds of things. You can be successful in life if you are strong or fast or—but you can also be successful in some evolutionary environments by being smart, and so I think there are going to be places in the universe where it is advantageous to be smart.

Mr. SCHWEIKERT. But the—I guess and—for Dr.—the fun in this one is how would you ever calculate it? Where—how would you ever sort of build your baseline to build from? And when you move from sort of hope, which is a powerful thing, to being able to put it into a calculator——

Mr. WERTHIMER. Yeah.

Mr. SCHWEIKERT. —there is often quite a leap there.

Mr. WERTHIMER. I think it is very difficult to estimate because we just have this one example on Earth. And so the—I think the only way we are going to find out is to do this search.

Dr. SHOSTAK. It is very akin, I think, to sitting around in the bars of Europe in 1700 trying to estimate the probability that any expeditions sent into the deep south—any sailing expedition will find the hypothesized southern continent there.

Mr. SCHWEIKERT. Yeah.

Dr. SHOSTAK. You know, what is the probability? Can you give me that to three figures before I fund to you? You can't. You can't.

Mr. SCHWEIKERT. Yeah. So——

Dr. SHOSTAK. You have to do the experiment.

Mr. SCHWEIKERT. So therefore it becomes a leap of faith but it is——

Dr. SHOSTAK. It is a reasonable leap of faith. It is a reasonable hypothesis that there is life to be found out there, even intelligent life to be found out there. And we can sit around and have a lot of drinks and talk about it, but in the end, if you don't do the experiment, you will just continue to have the drinks.

Mr. SCHWEIKERT. Well, seeing some of our questions, there may have been a lot of drinks going on.

Thank you, Mr. Chairman.

Chairman SMITH. Thank you, Mr. Schweikert.

Dr. Shostak, Mr. Werthimer, thank you both for your testimony, which was clearly appreciated by both Members of Congress as well as the audience.

And I also want to thank the Herndon High School students for being here today. You had a wonderful opportunity today to hear about a fascinating subject and I hope this will spur you on to study not only astrobiology but other scientific subjects as well.

And in case someone has an interest or wants to follow up on this subject, you might go to our Committee's website, which is Science.House.Gov, and we will clearly have some information about this hearing on that website, as well as other things that might be of interest to you all.

So thanks again for a wonderful hearing today and we stand adjourned.

[Whereupon, at 11:03 a.m., the Committee was adjourned.]

Appendix I

ANSWERS TO POST-HEARING QUESTIONS

ANSWERS TO POST-HEARING QUESTIONS

Responses by Dr. Seth Shostak

HOUSE COMMITTEE ON SCIENCE, SPACE, AND TECHNOLOGY

"Astrobiology and the Search for Life in the Universe"

Questions for the record, Dr. Seth Shostak, Senior Astronomer, SETI Institute

Questions submitted by Rep. Lamar Smith, Chairman, Committee on Science, Space, and Technology

1. SETI is an astrobiology program that uses the techniques and technology of astronomy. What are the unique challenges facing SETI research that might not be found in other areas of astrobiology?

   Most astrobiology exploration is focused on either finding habitats for life – such as exoplanets that might have the necessary conditions to support life – or finding life itself, either my direct exploration of the planets and moons of our solar system, or by spectrally analyzing the atmospheres of exoplanets in a search for biogenic gases.

   In both cases, astrobiologists know (a) where to search, and (b) what they are searching for (by analogy with terrestrial life).

   SETI differs from this in that very advanced intelligence might not even be biological, its location is not known, and the signals it could be emitting are also unknown. In other words, SETI scientists are required to use a far bigger net than the astrobiologists (and with far smaller budgets). In addition, SETI's instrumentation (radio telescopes, very fast photon detectors) are different than used in other astrobiology research.

   a. Is there a consensus in the SETI community about how to advance the science?

      Most SETI scientists would argue that the current level of effort is simply too small. With total expenditures on the order of a few million dollars a year (the majority from private funding) for all U.S. SETI projects combined, there are severe constraints on the breadth and depth of the reconnaissance, whether it be radio or optical SETI. The most effective way to advance the science would be to first ensure that the search can continue.

      Note that the efforts to find life (presumed microbial) on Mars – a popular line of astrobiology research – have budgets that are hundreds of times that of SETI.

2. At one point the United States was leading the world in the area of SETI research. Is this still the case?

   Yes, this is still true. While the U.S. efforts are very small, other countries are doing less. This is changing, however, with the construction of very large radio telescopes by the Europeans and the Chinese, both of whom have expressed a desire to do SETI.

3. You described in your written testimony that at the height of SETI's funding, it supported a two-pronged strategy: a low-sensitivity survey of the entire sky and a high-sensitivity targeted search of the nearest thousand star systems. Is this approach still in use today, or did it change when SETI lost federal funding?

    In some sense, this two-pronged strategy is still being followed, albeit somewhat inadvertently. For historic reasons, the SETI Institute's efforts are largely targeted searches, and it is now considering a reconnaissance of several tens of thousands of stellar systems using the Allen Telescope Array. The Berkeley optical SETI program is also targeted. The Berkeley radio searches, on the other hand, using the Arecibo telescope, are more in the vein of a sky survey – the telescope sweeps across the sky, and over the course of about two years, approximately one-third of the cosmos can be examined. Unfortunately, and unlike the NASA SETI project that was cancelled in 1993, this sky survey cannot quickly follow up on possible detections, and cannot dwell on any part of the sky for more than about 1-1/2 seconds.

    But with the above caveats, it's fair to say that SETI still follows a two-pronged strategy.

    a. If it's different, what are we losing out on by not having this two-pronged approach?

        As noted above, we are in fact using a two-pronged approach, although both the targeted search and the sky survey are not as robust as would have been the case with the NASA program.

    b. Is it possible that, by using only one approach, we could be missing information (for example, if a low-sensitivity survey doesn't pick up signals that the high sensitivity one does, and we're now only able to use the low-sensitivity survey).

        As both approaches are being followed, at least to some degree, this is not an issue.

    c. What would your survey approach be with unlimited funding?

        With more funding, better receivers and spectral analyzers could be built and used. This would increase sensitivity to weak signals, but the greater gain would be in the speed of the search. It would be possible to examine millions of star systems in the next two decades if support were available.

4. You said that SETI experiments are about 100 trillion times more effective today in terms of speed, sensitivity, and range of radio frequencies than in the 1960s. Are areas that were examined early on, before such sensitive equipment was developed, resurveyed? If not, couldn't we have missed signs of life?

In general, both targeted searches and sky surveys re-examine areas of sky that were the subject of previous searches. They do this with increased sensitivity, wider frequency coverage, or some other increment in effectiveness.

5. In our December 2013 hearing on Astrobiology, Dr. Steven Dick stated that he believed termination of the NASA SETI program in 1993 was based on the "'ridicule factor'" before planets were known to exist in abundance around Sun-like stars." Do you think this played a role in SETI's termination?

My response is, of course, speculative, but the so-called "giggle factor" was a greater concern among SETI researchers at the time of the 1993 termination than it is today. The detection of exoplanets that began in 1995 has, in my opinion, shifted the public's perception of the prospects for SETI. It is now generally perceived as well-motivated.

   a. If so, do you think that attitude has shifted with discoveries of exoplanets and even Earth's cousin (Kepler 186-f)?

   Yes, as described above. The giggle factor seems largely an attitude of the past.

   b. If not, what do you think was the biggest factor in SETI's termination?

   N/A

   c. What do you think is the biggest obstacle to broad public support for the SETI program to receive federal funding?

   While this is my perception only, I think that the public would support a modest, federally supported SETI program. The enormous interest in Mars exploration is strongly motivated by the possibility of life on that planet. The idea of "aliens" has been stoked by the entertainment industry and the popular-with-the-public idea that aliens might be visiting Earth (the UFO hypothesis). While few scientists are convinced by the evidence presented for visitation, there is no doubt a widespread fascination with, and acceptance of, the idea of intelligent extraterrestrial life.

   d. What is your most compelling argument for Congress to start funding SETI again, and what could this take the place of in NASA's current portfolio?

   A great deal of public money is spent on astrobiology – directly, in the form of robotic probes and research grants to scientists searching for life in our solar system -- and indirectly, in the funding of studies that try to find and characterize extrasolar planets. Hunting for signs of life beyond Earth is one of NASA's most visible activities, and indeed is perhaps second only to manned spaceflight in terms of public interest. And yet the most interesting possible discovery -- intelligent life – is not a steadfast part of the agency's agenda. This despite the fact that it would probably be the least expensive of all astrobiology research. A very substantial SETI program could be implemented at a cost of 0.1 percent or less of the total agency budget. The difficulty is not with the money, but with the allocation.

6. In the event that an intelligent civilization in the universe were to contact us, how would we understand the message that may be sent to us?

   There is no guarantee that we would understand any such message. The chances are improved if the signal turns out to be intended to be found by other societies. In that case, the senders would likely encode their information in a way that could be eventually understood by any society with radio receiving technology. One simple way to do this would be to transmit a series of bit map images, perhaps even a picture dictionary that would allow the recipient to eventually understand language.

   In any case, it's worth remarking that even if no information is garnered from a signal, the mere fact that there is other intelligence in the universe is important and profound.

7. The Transiting Exoplanet Survey Satellite (TESS) spacecraft, scheduled to launch in 2017, is designed to search for small exoplanets around the brightest stars near the Earth using the transit method. What benefits will TESS bring to SETI research?

   SETI benefits from every new exoplanet discovery, as this information allows SETI researchers to better hone their list of targets. In other words, these help in deciding where to aim our telescopes to have the best chance of examining a star system that's likely to have life, and even intelligent life. TESS will provide an inventory of relatively nearby exoplanets – and will lead to a fuller understanding of whether these worlds have atmospheres, and possibly even atmospheres that, by their composition, suggest the presence of biology. If you're going to hunt for life in the desert, it's good to know where the oases are.

8. What happens if astronomers make first contact with intelligent beings? Could you please describe the SETI Post Detection Protocols in place to handle that situation?

   The protocols are simply a "gentlemen's agreement" among some of the SETI groups world-wide about what to do in case a signal is found. Recent efforts by the SETI Permanent Committee of the International Academy of Astronautics have streamlined and simplified these protocols. However, they are actually quite simple: (1) Verify that any signal is truly an extraterrestrial transmission; (2) Inform the scientific community and the public; and (3) Refrain from transmitting any reply until there is international agreement on desirability and content.

   These protocols simply codify good research practice. They have no force of law.

9. In your interactions with the general public over the years, have you seen an increase or decrease in the number of people believing that we'll find intelligent life in the universe. To what do you attribute this?

I don't think there's been a very marked change. The public is, and has been for roughly a half-century, largely of the opinion that intelligent, extraterrestrial life exists. In the past, this opinion (which reaches as high as 80 percent of the population in some polls) was largely the result of the relentless appearance of aliens in popular media. However, during the past two decades, the steady drip of exoplanet discoveries – more and more of which involve planets that could sport life – has reinforced the idea that the cosmos could be replete with intelligent beings.

10. How many countries have SETI programs? Do any receive government funds?

Other than the U.S., the only country today with a continuing SETI program is Italy. Radio observations are made at the Medicina radio observatory of the University of Bologna. This is a university-sponsored effort, and therefore receives government funds. Other countries are contemplating SETI efforts, and these include several in Europe and Asia.

11. In your written testimony, you mentioned Asian and European radio astronomy initiatives, such as The Square Kilometer Array in Australia and South Africa. If it is completed, it will be the world's largest radio telescope. Why is the US not a participant in this project? Should it be?

The US had been involved in the SKA, through a university based consortium known as the U.S. SKA. The SETI Institute's Jill Tarter was chair of that consortium, and signed for US participation on the first SKA International Collaboration Agreement in 2000. That collaboration led to an NSF-funded technology development program resulting in the design approach for the SKA now being followed – a large number of small antennas, similar to the ATA.

According to Tarter, "when the SKA began to require the investment of "real" money to participate, asking for countries to buy a seat at the table, the NSF participated for a year or two as an advisor only and then backed out all together when a non-profit SKA organization was incorporated with member countries in 2011. The NSF argued that since the SKA was not highly recommended in the last U.S. Decadal Survey, there will be no money available for participation until at least the decade beginning in 2020. Whether the US can get back into the game at an affordable cost at that time is not yet clear."

a. Will the Square Kilometer Array be available for radio SETI use?

My understanding is that it will. There have been expressions of interest by European astronomers involved with the project. It would not be difficult to set up a SETI "feed" from the instrument for observations that are conducted simultaneously with other astronomical research. However, dedicated use of the SKA for specific SETI targets would likely require competitive proposals by researchers.

    b. Due to SETI's limited funding, has there ever been international cooperation with these initiatives specifically for SETI research? Why, or why not?

        See above.

    c. Is that something you would look to in the future? Why, or why not?

        There is simply no doubt that the SKA could be an extraordinarily powerful instrument for radio SETI. Of course researchers in this country would be interested in using it for SETI, if there were support monies that allowed them to do so.

12. How does SETI influence STEM education efforts?

The SETI Institute is, as an institution, very active in both formal and informal education projects. These range from teacher training to producing classroom materials. I have co-authored a college textbook on astrobiology.

But aside from these deliberate efforts, the idea of extraterrestrial life, and most notably intelligent extraterrestrial life, is an effective hook in getting young people interested in the science of astrobiology. It motivates them to study biology, geology, astronomy and planetary physics.

13. How many universities include SETI research in their astrobiology course selections?

    While I am not able to speak with precision to this question, I know that 141 colleges and universities use the "Life in the Universe" textbook that Jeffrey Bennett and I have authored. According to the publishers (Pearson Education), our text has been adopted by more than 70 – 80 percent of the undergraduate courses in astrobiology offered by such schools, so this tally gives an indication of how widespread these courses are in the U.S. Note that it has an entire chapter devoted to SETI, and other chapters cover related topics (e.g., exoplanets).

Questions for the Record from
Ranking Member Johnson
Full Committee Hearing
"Astrobiology and the Search for Life in the Universe"
May 21, 2014

Dr. Seth Shostak

1. Please elaborate on the relationships between astrobiology, exoplanet discoveries, and the search for extraterrestrial intelligence. In addition,
   - Are there any astrobiology findings or discoveries of Earth-size and Earth-like planets that would significantly influence the SETI project and how it is carried out?

The discovery of thousands of exoplanets during the last two decades has profoundly influenced SETI research in two ways: (1) It has told us which stars have planets, and of greater importance, which fraction of stars have planets. More recently, this work has uncovered some planets that are in their star's habitable zone, and therefore good candidates for having life.

In other words, astrobiology research is improving the quality of SETI "targets" by telling us the best places to aim our telescopes. (2) Of perhaps greater import, these studies have shown that nearly all stars have planets, and perhaps as many as one in five may have a "cousin of the Earth" in orbit about it. And finally, (3) studies of our own solar system, as well as of extremophiles here on Earth, have revealed that moons might be as amenable to life as planets, and that life can exist in very difficult environments. All of these results point in the same direction: There is more reason to suspect that life is common in the cosmos.

   - Are there results from ongoing NASA missions that are particularly useful to the SETI effort, and are there planned NASA missions that you anticipate will help inform SETI research? If so, why?

The most obviously useful NASA missions for SETI are (1) Telescopes that can find and/or characterize planets (and eventually large moons) around other stars. These allow SETI to narrow its searches. (2) The continuing efforts to find evidence that Mars once had (or perhaps still has) microbial life, as well as future plans to do the same for some of the moons of the outer solar system, such as Europa, Enceladus, or Titan. Finding biology on any of these worlds would give scientists a second "data point" as to the prevalence of life. We think biology is a commonplace phenomenon, but success with any of these projects would turn speculation into fact.

2. In 2011, the Air Force initiated a formal assessment of the Allen Telescope Array's (ATA) utility for Space Situational Awareness.
   - How would the Air Force Space Situational Awareness requirements impact the daily use of the ATA for SETI purposes?

There is an up side and a down side. The benefits are that the Air Force (though their contracts with SRI International) pays for the operation and maintenance of the ATA, thus removing this financial burden from the cash-strapped SETI Institute. The disadvantage is that the time available for SETI observations is approximately half what it otherwise would be.

- What were the results of the Air Force assessment and what is the current status of the Air Force's use of the ATA?

The Air Force continues to use the ATA for its SSA work, primarily as a test bed. While there is no guarantee that they will continue to do so beyond the current contracts with SRI, they seem to be satisfied with the efforts so far.

3. Much larger radio arrays replicating the ATA concept are under construction in a phased development timeline as part of the Square Kilometre Array (SKA) Project. The Square Kilometre Array is a global science and engineering project which envisions building the world's largest radio telescope using hundreds of thousands of radio telescopes. Although SKA precursors are planned, the eventual SKA telescopes will be co-located in Africa and in Australia.
    - The last Decadal Survey on Astronomy and Astrophysics characterized the ATA as one of several SKA-mid (3 to 100-centimeter wavelength) pathfinder instruments. Is there a formal relationship between ATA and SKA and if so, what are the key features of this relationship?

See the answer regarding the SKA above. There is no formal relationship between the two projects. The ATA, however, has been a leader in pioneering the paradigm of building radio telescope arrays with large numbers of small individual antennas (the so-called "LNSD" design). This approach has many advantages for research, particularly for making radio images of the sky and for quickly surveying large tracts of the cosmos. In the past, it was not feasible to build LNSD arrays, simply because of the very high cost of the electronics required for each of the antennas. However, in the last 20 years or so, the price of the requisite electronics has dropped by roughly a factor of 100, and changed the entire schema of such instruments. The SKA owes much of its design philosophy to the ATA.

- How suitable would the SKA be for the search for extraterrestrial intelligence? Was that potential use a major driver for SKA, or were other applications the main focus?

  See question 11 above. While SETI is certainly not the major driver for building the SKA – any more than SETI or radio astronomy of galaxies were the main drivers for constructing the Arecibo radio telescope (even those are two projects that have often been on the telescope since its beginnings) – the tremendous sensitivity of the SKA will make it a very attractive instrument for SETI. The SKA should be thought of as a general purpose radio astronomy tool. It will be used for a wide range of research projects. SETI will be among them.

- What is the impact of SKA on the future outlook for the ATA?

  The ATA cannot compete with the SKA on the basis of raw sensitivity. The latter will have far larger collecting area, and can also reach areas of the southern sky that the ATA simply can't observe. But the ATA can dedicate a very substantial fraction of its research program to targeted SETI searches. That's something that the general purpose SKA will not be able to do. In astronomy, no one telescope can ever "do it all." It's also worth noting that the SKA doesn't exist now, and might not be operational for 10 – 15 years more.

- Are there plans for an expansion of the ATA beyond 42 dishes, and if so, what are the plans and how would they be funded?

  The original design goal for the ATA was to build 350 antennas. This didn't happen, simply because of a lack of funding. It would greatly increase both the sensitivity and the flexibility of the instrument to build it out to at least 200 dishes. The cost of doing so is estimated to be about $35 million. Raising that money from private donors is being tried, but has so far been unsuccessful.

4. What are the key educational opportunities associated with the search for extraterrestrial intelligence?

   SETI involves a wide range of science disciplines, from astronomy to geology and biology (and even the social sciences, if one considers questions about the societal reaction to a detection). It is an essential part of any course on astrobiology. But more than that, the tremendous public appeal of the SETI enterprise serves to interest young people in the subject areas of astrobiology. I give approximately 50 talks annually, most of them at colleges and universities. At virtually every one of these presentations, I am approached by students who would like to pursue careers in this field, work for SETI, or both. It is wonderful hook to get people interested in science.

Questions for the Record

Zoe Lofgren
Astrobiology and the Search for Life in the Universe

I am sorry to have been unable to attend this important hearing in the Science Committee. I was particularly disappointed, as both witnesses are from the Bay Area.

Questions for Dr. Shostak:

As one of SETI Institute's stated objectives is to engage the public through education and public outreach, are you familiar with the Lick observatory, and have you found it valuable or do you think it could be valuable in that public education and outreach goal?

While I have not personally used the Lick Observatory (I am a radio astronomer), the SETI Institute has in the past conducted a so-called "optical SETI" experiment using the 1 meter Nickel telescope at Lick. The experiment, which used equipment specially built by a UC Santa Cruz undergraduate, searched star systems for brief, flashing laser pulses that would betray the presence of intelligence. There have been serious plans to once again use Lick for this type of work, but these have not yet been funded.

While I cannot attest to the research value of Lick, there is little doubt in my mind that it is a very valuable site for exposing Bay Area residents – particularly young people – to professional astronomical research. But continuing to reap this benefit will require that at least some of the research programs there continue to be funded. Once lost, they cannot be regained.

Along those same lines, the Automated Planet Finder has only recently come on line at Lick. Are you familiar with this tool? Might it be useful in SETI Institute's research into finding habitable planets?

UC's Automated Planet Finder is, of course, a valuable tool in extending our knowledge of exoplanets ... and those bear directly on SETI. The more we know about possible habitats for life, the greater the chance that we will be successful in finding it.

Responses by Mr. Dan Werthimer
SST Full Committee Hearing
"Astrobiology and the Search for Life in the Universe"
May 21, 2014
Elizabeth H. Esty Questions for the Record

Thank you, Chairman Smith, for holding this hearing, and thank you to the witnesses for your testimony about the current state of the search for life in the universe.

Drs. Shostak and Werthimer, you both spoke about the incredible amount of data you—and other SETI researchers—analyze every day. I am particularly interested in how these efforts can be expanded through crowdsourcing. Specifically, I want to know more about the work you are doing, or the work that could be done, to engage students at the elementary and secondary school levels to help with culling this data. Could you explain further the role you see for students—and citizen scientists at home--to assist in these efforts?

There have been several initiatives to make SETI data available to the general public, both as a means of increasing the scrutiny of these data as well as engaging the populace in the excitement of this effort. The best-known of these is UC Berkeley's SETI@home screen saver, which I believe has been running for close to a decade now and was described by Dan Werthimer. The SETI Institute has also had an interactive data analysis project, called SETIlive. However, because of the manner in which observations are made at the Allen Telescope Array, citizen processing of these data is of lesser use than for the Berkeley projects. In particular, because the ATA can be specifically directed at a target star system, it is able to follow up immediately on any likely looking signal. That is obviously a desideratum, given that there's no guarantee that a signal coming our way will be persistent – it might go away after minutes, hours, or days. Consequently, the ATA follows immediately on any candidate signals, something that is far harder with a citizen science approach.

As "big data" becomes more tractable for processing by smart phone apps and other modes of ingress by the public, I have little doubt that other SETI data processing schemes will be launched. The value of this is probably less in the research potential than in the benefits of public participation in the project.

UNIVERSITY OF CALIFORNIA, BERKELEY

Dan Werthimer
Director, Berkeley SETI Research Center
http://seti.berkeley.edu

SPACE SCIENCES LABORATORY
7 Gauss Way
Berkeley, CA 94720-7450

4 July 2014

Dear Mr. Chairman and Members of the House Committee on Science, Space and Technology,

Thank you for the opportunity to testify last month on "Astrobiology and the Search for Life in the Universe", and thank you for your thoughtful follow-up questions. I hope you find these answers useful:

Questions submitted by Chairman Lamar Smith:

1. At one point the United States was leading the world in the area of SETI research. Is this still the case?

The United States continues to lead SETI research; about 2/3 of the world's SETI research is in the USA. Currently, the best telescopes on our planet for SETI observations are operated by NSF and NASA: the world's largest radio telescope, in Arecibo, Puerto Rico, and the world's largest fully steerable radio telescope, the Robert C. Byrd Telescope in Green Bank, West Virginia. However, NSF plans to cut all funding for the Green Bank Observatory, and NSF has significantly reduced funding for Arecibo. China is building larger telescopes than both the Green Bank and the Arecibo telescopes. And many countries are working together to construct giant telescope arrays in South Africa and Australia; the US is not currently participating in this international "Square Kilometer Array" project. In a few years, the best places for SETI observations will likely be in China, South Africa, and Australia.

2. In your written testimony, you explained the "panchromatic SETI" program, which will search nearby stars and those determined to be most likely to host a solar-like exoplanet system. What criteria do you use to determine stars likely to host a solar system?
a. A March 25, 2014, article in the *International Journal of Astrobiology* ("Habitability around F-type stars" by Sato, Cuntz, Guerra Olvera, Jack, and Schroder) urges astronomers to consider larger F-type stars in the search for extraterrestrial life. Has this or other new research influenced your criteria?

In several of our SETI projects, we target sun-like stars with planets similar to Earth (planets similar in size, that are in the "habitable zone" or "Goldilocks zone" - not too hot, not too cold). However, as your question points out, this targeting strategy might be too anthropocentric; life may thrive and evolve on planets and stellar systems that are very different from ours. Our Panchromatic SETI project targets stars using many techniques and using many observatories operating in different parts of the electromagnetic spectrum. To choose the targets in this project, we are use a less biased scheme: we target all nearby stars, independent of whether these stars are similar to our sun, or whether these stars are known to have planets.

In the last year, the number of known and candidate exoplanets has reached a level that allows a rich statistical analysis of the correlations between stellar properties and their planet populations: relationships that may make it possible to infer the likelihood that a particular star might host an unseen planet like the Earth. Identifying such analogs individually, with certainty, may take years, but probabilistic statements can be made now. Stellar mass, composition, and known exoplanets have been shown to be predictive of the aggregate properties of a star's complete exoplanet system.

The idea of a "habitable zone," originally defined as the region around a star where the temperature was such that water could exist in a liquid state, has now given rise to a panoply of research to assess the likelihood or suitability for life in extraterrestrial environments. In general, this body of work indicates that a very wide range of environments could be suitable for life. While we constantly assess such work and collaborate closely with our colleagues to target the most interesting stars and planetary systems, we temper our confidence in our understanding of habitability by also conducting more agnostic searches, such as surveys of large fractions of the entire sky or targeted observations of stars based only on their proximity to the Earth.

3. In our December 2013 hearing on Astrobiology, Dr. Steven Dick stated that he believed termination of the NASA SETI program in 1993 was based on the "'ridicule factor'" before planets were known to exist in abundance around Sun-like stars." Do you think this played a role in SETI's termination?
   a. If so, do you think that attitude has shifted with discoveries of exoplanets and even Earth's cousin (Kepler 186-f)?
   b. If not, what do you think was the biggest factor in SETI's termination?
   c. What do you think is the biggest obstacle to broad public support for the SETI program to receive federal funding?
   d. What is your most compelling argument for Congress to start funding SETI again, and what could this take the place of in NASA's current portfolio?

I don't know what was in Senator Bryan's mind when he led the effort to terminate NASA's SETI program. People who have researched this suggest that Senator Bryan was looking for a federally supported program which could be ridiculed and misconstrued as a quack project looking for "little green men", and which would attract public attention so that he could present himself as a protector of the tax-payer's money. Senator Bryan had attempted, and failed, to cut SETI before; he succeeded in terminating NASA's program at a session of the House where very few members of congress were present.

Since then, as your question points out, exciting discoveries have been made about extra-solar planets and the potential for life in the universe. There has always been strong public support for SETI, and now, because of these new discoveries, it's even stronger.

SETI is fragile and its future is uncertain - there are only a handful of people conducting SETI experiments in the USA and many SETI researchers are nearing retirement. We hope in the future, that NSF and NASA can fund a rich variety of small scale and perhaps eventually mid-scale peer reviewed SETI projects at several different institutions, especially projects that train new students and postdoctoral fellows. This would make SETI more robust, encourage new innovative ideas and search strategies, and could eventually lead to the detection of life and civilizations beyond Earth.

4. In 2018, NASA will launch the James Webb Space Telescope (JWST). The telescope's infrared ability will enable scientists to learn more about exoplanets targeted by TESS. How will this new capability affect SETI research?

5. How will the wide-field sky surveys conducted by the future Large Synoptic Survey Telescope (LSST) assist optical SETI research?

The flood of multi-planet systems discovered by the Kepler mission and the high precision of the planetary orbital data has directly made many of our most recent targeted searches possible. We're very excited about TESS (the Transiting Exoplanet Survey Satellite). TESS will find multi-planet systems that are much closer to the Earth than the Kepler exo-planets. TESS will produce a great target list for our "Eavesdropping SETI" project.

There are several ways in which large upcoming astronomical observatories could affect SETI research. They will detect new individual extra-solar planet systems and shed light on our understanding of exo-planet formation, which will lead to the refinement of the targeted catalogs used in SETI searches. One of our basic strategies in SETI is to target environments as similar as possible to the one we know was successful at giving rise to intelligent life - our own Solar System. JWST will be able to analyze the composition of exo-planet atmospheres, by conducting infra-red spectroscopy. More generally, TESS, JWST and LSST will all play a role in improving our understanding of how our sun and our Solar System fit in to our picture of the Milky Way.

Another more speculative possibility is that, by studying the sky with better instruments and in different ways, these new observatories may discover something completely unexpected that will have bearing on the search for intelligent life. There is a long history of serendipity playing a critical role in new astronomical discoveries. In dedicated SETI experiments, we conduct the most methodical searches we can, using the best techniques we can think of, but the detection of extraterrestrial intelligent life could come from a careful scientist doing an experiment having nothing to do with SETI.

6. In your interactions with the general public over the years, have you seen an increase or decrease in the number of people believing that we'll find intelligent life in the universe. To what do you attribute this?

I'm not an expert in public perception of life in the universe, but I think more and more of the public are excited about SETI and the possibility of life elsewhere. A part of this interest stems directly from science (eg: the discovery of extra-solar planets and water on moons in our own solar system), and part stems from literature and film. The book and subsequent popular movie Contact, were written by Carl Sagan, an expert in astrobiology and SETI. Five hundred million people watched and were inspired by Sagan's COSMOS series. Several million people have downloaded the SETI@home screen saver program and are helping us hunt for ET.

7. How many countries have SETI programs? Do any receive government funds?

There have been SETI projects in the USA, Netherlands, Italy, Australia, England, France, Japan, Russia, and Argentina (this list of countries might not be complete). Others are interested and hoping to launch small SETI projects in the next few years. I think most SETI projects in other countries receive small government funding, mostly in the form of salary for researchers at government labs and universities. There are currently about two dozen people on the planet who conduct SETI observations.

8. How does SETI influence STEM education efforts?

SETI@home has engaged millions of volunteers in a global science project. Thousands of students are participating in SETI@home, and SETI@home is part of the Great Explorations in Math and Science (GEMS) curriculum. There is a lot of wonderful curriculum that explores the possibility of life in the universe. The question "Are we alone?" is fascinating to many young children, and touches on many disciplines, including physics, astronomy, chemistry, biology, engineering, math, and computer science, and is a great way to get kids engaged in science.

9. How many universities include SETI research in their astrobiology course selections?

Many universities offer an introductory course about life in the universe. This is often a popular course for non-science majors, as it touches on so many areas in science (see question 8). There are a handful of universities that offer graduate courses and student research opportunities in astrobiology.

Questions from Ranking Member Johnson

1. Please elaborate on the relationships between astrobiology, exoplanet discoveries, and the search for extraterrestrial intelligence. In addition, Are there any astrobiology findings or discoveries of Earth-size and Earth-like planets that would significantly influence the SETI project and how it is carried out? If so, why? Are there results from ongoing NASA missions that are particularly useful to the SETI effort, and are there planned NASA missions that you anticipate will help inform SETI research? If so, why?

Please see my answers to Chairman Smith's questions 4 and 5 about existing NASA missions. NASA's Kepler mission has been spectacular, discovering thousands of extrasolar planets, and the upcoming TESS mission will provide key information about planets on nearby star systems. Missions to study Europa and Enceladus would be tremendously exciting, as these moons have liquid water and perhaps harbor primitive life. More generally, manned space missions have little or no science and are extremely expensive. Robotic space missions are less expensive, less dangerous, and provide considerably more scientific knowledge.

2. During the question and answer portion of the hearing, you provided some explanation in response to a question on the security of the SETI@Home software. That question related to a study in which you and your team found that there had been two noteworthy attacks and that the web server for SETI@Home was compromised. I understand that you also found later that hackers stole thousands of user email addresses by exploiting a design flaw in your client/server protocol. Please provide details on the current state of the security of the SETI@Home software. Have there been further assessments, including independent assessments, of the security of the SETI@Home software? If not, why not?

The attacks on the SETI@home server exploited vulnerabilities in system software (operating system, web server), not in the SETI@home software. To deter further attacks we follow security best-practices on our servers, such as applying the latest security patches for all system software.

The SETI@home software is based on BOINC, a system for volunteer computing that is also used by about 50 other science projects. The basic design of BOINC emphasizes security. For example, BOINC uses encryption-based "code signing" to ensure that even if hackers break into a project's server, they will not be able to distribute malware via BOINC.

BOINC has been subject to several independent security reviews.

1) IBM Corporation uses BOINC as part of a philanthropic project called "World Community Grid", and also runs BOINC on thousands of their in-house computers. They have done several security reviews of BOINC, the first in 2005, the most recent in 2013.

2) Intel Corporation operates a project called "Progress Thru Processors" that promotes BOINC through a Facebook interface, and they run BOINC on in-house computers. They did a security review of BOINC in 2009.

3) HTC Corporation (from Taiwan) operates a project called "HTC Power to Give" that promotes BOINC running on Android smart phones. They did a security review of BOINC in 2014.

4) CERN (the European physics center) did a security analysis of BOINC's web code in 2013.

Each of these reviews identified several potential vulnerabilities, all of which are now fixed.

Because the BOINC source code is open-source, it is constantly being studied by volunteer programmers, who occasionally submit security-related improvements.

BOINC uses several open-source libraries, such as the OpenSSL security library. When security vulnerabilities are fixed in these libraries (such as the recent Heartbleed vulnerability in OpenSSL) BOINC immediately incorporates these fixes.

3. What are the key educational opportunities associated with the search for extraterrestrial intelligence?

Please see my answers to Chairman Smith's questions 8 and 9 about educational opportunities.

Questions from Zoe Lofgren

I am sorry to have been unable to attend this important hearing in the Science Committee. I was particularly disappointed, as both witnesses are from the Bay Area.

Questions for Dr. Werthimer:
In your testimony you mention SETI research being done at Lick Observatory (which is located on Mt. Hamilton in my district). Could you discuss the importance and utility of Lick, and whether you share my concern that the University of California is planning to cease funding this facility? As an astronomy professor at Berkeley, is it useful to you and your students to have a major astronomical observatory in the state of California, rather than only in Hawaii or farther afield?

The University of California and other California educational institutions have suffered severe budget cuts since 2008. In my opinion, these cuts are short sighted, and I hope can be corrected soon. Education, especially science education, is crucial to good jobs, a health economy, and a healthy democracy.

As you point out, UC is planning to cease funding for Lick Observatory in 2018, as a result of these cuts.

We have conducted novel SETI research at UC's Lick observatory, and we are hoping to launch a new SETI search there later this year, searching for nanosecond infrared light pulses that might be emitted by alien civilizations, something no one has ever tried before.

This kind of work, developing and testing new instruments and ideas, cannot be done on large telescopes (such as Keck in Hawaii) because there simply isn't enough observing time available for such risky, exploratory projects -- and such large telescopes would be overkill. We can develop the technology at the Lick Nickel telescope and use it to demonstrate that the technique is able to potentially detect weak signals. We don't have any telescope for this development work available to us elsewhere. Another example of instrument development at Lick is the instrument that Professor Geoff Marcy and colleagues pioneered to detect the first extrasolar planets. Lick continues to be used by Marcy and others for exoplanet searching and exoplanet charaterization.

Moreover, there is considerable student involvement in our projects because Lick is so nearby. Indeed, our early optical SETI system was largely developed by Shelly Wright, then a undergraduate student, and now a professor at the University of Toronto. This is a perfect project for student training, and such involvement is a tremendous strength of Lick Observatory.

Losing Lick observatory would be a huge loss for UC research and education.

Question from Elizabeth H. Esty

Drs. Shostak and Werthimer, you both spoke about the incredible amount of data you—and other SETI researchers—analyze every day. I am particularly interested in how these efforts can be expanded through crowdsourcing. Specifically, I want to know more about the work you are doing, or the work that could be done, to engage students at the elementary and secondary school levels to help with culling this data. Could you explain further the role you see for students—and citizen scientists at home--to assist in these efforts?

As I mentioned in my answer to Chairman Smith's question 8, the SETI@home citizen science project has engaged millions of volunteers in a global science project. Thousands of K - 12 students are participating in SETI@home, and SETI@home is part of the Great Explorations in Math and Science (GEMS) curriculum. The question "Are we alone?" is fascinating to many children, and touches on many disciplines, including physics, astronomy, chemistry, biology, engineering, math, and computer science. It is a great way to get kids engaged in science.

Our SETI group developed general purpose open source software for public participation volunteer computing, called BOINC (Berkeley Open Infrastucture for Network Computing). Using this software, millions of volunteers are participating in 50 scientific computing projects, including malaria, cancer, and HIV drug research, climate modeling, pulsar searching, protein folding and SETI. Participants can use their spare computing cycles for the science projects they are most interested in. Some have called this the democratization of scientific supercomputing.

We also developed the Stardust@home project with our Berkeley colleagues, where tens of thousands of volunteers use their brains, eyes, and a "virtual microscope" program on their home computer to find microscopic cometary and interplanetary dust particles from NASA's stardust mission.

We are working on developing a new citizen science project, dubbed "Energy at Home" (currently unfunded), where we plan to ask kids to conduct an "energy audit" of their home, using meters to measure the energy usage of electrical appliances - refrigerators, TV's, computers, phones... comparing each other's homes and appliances, and hopefully finding new ways to conserve energy.

Best Wishes,

Dan Werthimer
Director, Berkeley SETI Research Center
University of California, Berkeley